Röbbe Wünschiers

Computational Biology -
Unix/Linux, Data Processing and Programming

Springer

Berlin
Heidelberg
New York
Hong Kong
London
Milan
Paris
Tokyo

Röbbe Wünschiers

Computational Biology – Unix/Linux, Data Processing and Programming

With 19 Figures and 12 Tables

Dr. Röbbe Wünschiers
University of Cologne
Institute for Genetics
Weyertal 121
50931 Köln
Germany

ISBN 3-540-21142-X Springer-Verlag Berlin Heidelberg New York

Library of Congress Control Number: 2004102411

Springer-Verlag is a part of Springer Science + Business Media
springeronline.com

Springer-Verlag Berlin Heidelberg 2004
Printed in Germany

Cover design: Design & Production, Heidelberg
Typesetting: Camera ready by the author
31/3150WI - 5 4 3 2 1 0 - Printed on acid-free paper

Dedicated
to
Károly Nagy and to the Open-Source Community

Foreword

A shift in culture

Only a decade ago, the first thing a molecular biologist would have had to learn when he or she started the lab work was how to handle pipettes, extract DNA, use enzymes and clone a gene. Now, the first thing that he or she should learn is how to handle databases and to extract all the information that is already known about the gene that he or she wants to study. In all likelihood, he or she will find that the gene has already been sequenced from several organisms, that it was recovered in a variety of EST projects, that expression data are available from microarray and SAGE studies, that it was included in linkage studies, that proteomics data are rapidly accumulating, that lists of interacting proteins are being compiled, that domain structure data are available and that it is part of a network of genetic interactions which is intensively modelled. He or she will discover that all this information resides in many different databases with different data formats and with different levels of analyses and linking. Starting to work on this gene will make sense only, if all this information is put together in a project-specific manner and set into the context of what is known about related genes and processes. At this point he or she may decide to walk up to the bioinformatics group in house and ask for help with arranging the data in a useful manner. This will then turn into the first major frustration in his or her career, since the last thing a scientific bioinformatics group wants to do is to provide a service for data retrieval and management.

Molecular biology is currently going through a dramatic cultural shift. The daily business of pipetting and gel running has increasingly to be complemented with data compiling and processing. Large lists of data are produced by sequencers, microarray experiments or real-time PCR machines every day. Working with data lists has become as important as extracting DNA. Every bench scientist needs proficiency in computing; discoveries are made both at the bench and on the screen. It would be completely wrong to think that computers in molecular biology are the business of bioinformaticians only.

Bioinformatics has become a scientific discipline of its own and should not be considered to be a service provider. The day-to-day computing will always have to be done by the experimentalist himself or herself.

Of course, there are now also a lot of helpful and fancy program packages for the bench scientist; but these will only perform routine tasks and all too often they are only poorly compatible. A scientist needs the freedom to develop his or her own ideas and to link things that have previously not been linked. Being able to go back to the basics of computing and programming is therefore a vital skill for the experimentalist, as important as making buffers and setting up enzyme reactions. It allows him or her to handle and analyze the data in exactly the way it is required for the project and to pursue new avenues of research, rather than trotting old paths.

Unix is the key to basic computing. If one is used to Windows or Mac operating systems, this might at first sound like going back into the stone age; but the dramatic recent shift of at least the Mac operating system to a Unix base should teach us otherwise. Unix is here to stay and it allows the largest flexibility for bioinformatics applications. Those who have learned Unix will soon discover the myriad of little "progies" that are available from colleagues all over the world and that can make life much easier in the lab.

This book gives exactly the sort of introduction into Unix, Unix-based operating systems and programming languages that will be a key competence for experimentally working molecular biologists and that will make all the difference for the successful projects of the future. It has been written by a bench scientist, specifically with the needs of molecular biologists in mind. It can be used either for self-teaching or in practical courses. Every group leader should hand over this book to new students in the lab, together with their first set of pipettes.

Cologne/Germany,
January 2004

Prof. Dr. Diethard Tautz
(Institut für Genetik der Universität zu Köln)

Preface

Welcome on board!

With this book I would like to invite you, the scientist, to a journey through terminals and program codes. You are welcome to put aside your pipette, culture flask or rubber boots for a while, make yourself comfortable in front of a computer (do not forget your favourite hot alcohol-free drink) and learn some unixing and programming. *Why?* Because we are living in the information age and there is a huge amount of biological knowledge and databases out there. They contain information on almost everything: genes and genomes, rRNAs, enzymes, protein structures, DNA-microarray experiments, single organisms, ecological data, the tree of life and endless more. Furthermore, nowadays many research apparatuses are connected to computers. Thus, you have electronic access to your data. However, in order to cope with all this information you need some tools. This book will provide you with the skills to use these tools and to develop your own tools, i.e. it will introduce Unix and its derivatives (Linux, Mac OS X, CygWin, etc.) and programming (shell programming, awk, perl). These tools will make you independent of the way in which other people make you process your data – in the form of application software. What you want is open functionality. You want to decide how to process (e.g. analyze, format, save, correlate) data and you want it now – not waiting for the lab programmer to treat your request; and you know it best – you understand your data and your demands. This is what open functionality stands for, and both Linux and programming languages can provide it to you.

I started programming on a Casio PB-100 hand-held built in 1983. It can store 10 small Basic programs. The accompanying book was entitled "Learn as you go" and, indeed, in my opinion this is the best way to learn programming. My first contact to Unix was triggered by the need to copy data files from a Unix-driven Bruker EPR-Spectrometer onto a floppy disk. The real challenge started when I tried to import the files to a data-plotting program on the PC. While the first problem could be solved by finding the right page in a Unix manual, the latter required programming skills – Q-Basic at that time. This problem was minor compared to the trouble one encounters today. A common problem is to feed one program with the output of another program: you might have to change lines to columns, commas to dots, tabulators to semicolons, uppercase to lowercase, DNA to RNA, FASTA to GenBank format and so forth. Then there is that huge amount of information out there in the web, which you might need to bring into shape for your own analysis.

You and This Book – This book is written for the total beginner. You need not even to know what a computer is, though you should have access to one and find the power switch. The book is the result of a) the way how I learned to work with Unix, its derivatives and its numerous tools and b) a lecture which I started at the Institute for Genetics at the University of Cologne/Germany. Most programming examples are taken from biology; however, you need not be a biologist. Except for two or three examples, no biological knowledge is necessary. I have tried to illustrate almost everything practically with so-called terminals and examples. You should run these examples. Each chapter closes with some exercises. Brief solutions can be found at the end of the book.

Why Linux? – This book is not limited to Linux! All examples are valid for Unix or any Unix derivative like *Mac OS X*, *Knoppix* or the free Windows-based *CygWin* package, too. I chose Linux because it is open source software: you need not invest money except for the book itself. Furthermore, Linux provides all the great tools Unix provides. With Linux (as with all other Unix derivatives) you are close to your data. Via the command line you have immediate access to your files and can use either publicly available or your own designed tools to process these. With the aid of *pipes* you can construct your own data-processing pipeline. It is great.

Why `awk` and `perl`? – `awk` is a great language for both learning programming and treating large text-based data files (contrary to binary files). To 99% you will work with text-based files, be it data tables, genomes or species lists. Apart from being simple to learn and having a clear syntax, `awk` provides you with the possibility to construct your own commands. Thus, the language can grow with you as you grow with the language. I know bioinformatic professionals entirely focusing on `awk`. `perl` is much more powerful but also more unclear in its syntax (or flexible, to put it positively), but, since `awk` was one basis for developing `perl`, it is only a small step to go once you have learned `awk` – but a giant leap for your possibilities. You should take this step. By the way, both `awk` and `perl` run on all common operating systems.

Acknowledgements – Special thanks to Kristina Auerswald, Till Bayer, Benedikt Bosbach and Chris Voolstra for proofreading, and all the other students for encouraging me to bring these lines together.

Hürth/Germany,
January 2004 *Röbbe Wünschiers*

Contents

Part I

Whetting Your Appetite

1

Introduction

1.1 Information

This book aims at the total beginner. However, if you know something about computers but not about programming, the book will still be useful for you. After introducing the basics of how to work in the Unix/Linux environment, some great tools will be presented. Among these are the stream line editor `sed` and the script-oriented programming languages `awk` and `perl`. These utilities are extremely helpful when it comes to formatting and analyzing data files. After you have worked through all the chapters, you can use this book as a reference. The learning approach is absolutely practically oriented. Thus, you are invited to run all examples, printed in so-called Terminals, on your own!

If you face any problems: contact me! Of course, I cannot help you if your non-unix-like-operating-system driven computer crashes continuously. However, if things connected to this book confuse you – or you even find errors – please let me know:

Email: *rw@biowasserstoff.de*

Further information about this book, including lists with internet links and known errors, can be found at my homepage.

Homepage: *www.uni-koeln.de/~aei53*

You are very much welcome to supply me with good ideas for examples!

1.2 What Is Linux?

Linux is a multi-user multi-task operating system, originally based on *Mimix*, which is an operating system similar to Unix. Linux was initially developed by Linus Torvalds in 1991. It is an *open source* operating system. This means everybody who has programming knowledge can modify and improve the system; but it also means that everybody can download and install it. This is a main reason to choose Linux: you need invest no money except for the book itself. Still, the content of this book is valid for any Unix-like operating system like Mac OS X or CygWin, the free Unix emulator for Windows.

1.3 What Is Shell Programming?

The shell, also called terminal or console, will be our playground. Everything we do in this book is done in the shell. The shell can be seen as a command interpreter: we enter a command and the shell takes care of its execution; but we can also combine a number of commands and programs, including programming structures like decisions, in order to generate new functionality. Typical shell programs handle files and directories rather than file contents. A common task would be to convert all file extensions from *.txt* to *.seq*, make specific files executable or archive all recently changed files. Shell programming resembles DOS's programming language for *batch files*.

1.4 What Is `sed`?

`sed` (**s**tream **ed**itor) is used to perform basic text editing on an input text file (or data stream) and was written by Lee E. McMahon in 1973. `sed` does not allow for any interactions.

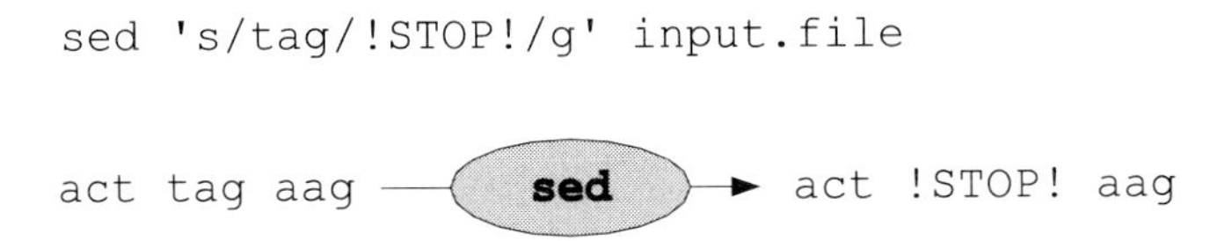

Fig. 1.1. What does `sed` do?

Figure 1.1 shows a sketch for an example of what `sed` can do. Here, the stop codon of a DNA sequence is replaced by the text "!STOP!". `sed` is well suited to perform small formatting tasks like converting RNA to DNA, commas to points, tabs to semicolons and the like.

1.5 What Is `awk`?

`awk` is a programming language for handling common data manipulation tasks with only a few lines of code. It was initially developed by Alfred V. **A**ho,

Peter J. **W**einberger and Brian W. **K**ernighan in 1977. This is where the name comes from. `awk` is really a great tool when it comes to analyzing the content of data files. With `awk` you can perform calculations, draw decisions, read and write multiple files. What is best is that `awk` can be extended with your own designed functions. A typical task would be to fuse the content files having one common field. Another typical task would be to extract data matching certain criteria as shown above. `awk` forms the kernel of this book. After you have finished the chapter on `awk` you should be able to a) program basically anything you need and b) learn any other programming language.

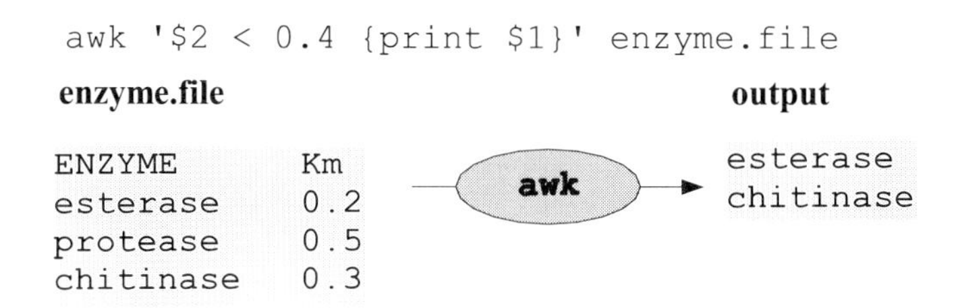

Fig. 1.2. What does `awk` do?

The example shown in Fig. 1.2 shows one basic function of `awk`. All enzyme names in the file *enzymes.file* are printed, if the corresponding Km value (do you remember enzyme kinetics and Michaelis-Menten?) is smaller than 0.4.

1.6 What Is `perl`?

`perl` (**p**ractical **e**xtraction and **r**eport **l**anguage) arose from a project started in 1987 by Larry Wall. It combines some of the best features of the programming language C, `sed`, `awk` and shell programming and was optimized for scanning arbitrary text files, extracting information from those text files and printing reports based on that information. `perl` is intended to be practical rather than being a beautiful language. It offers you everything a programming language can offer. `perl` is often used to program web-based applications (known as CGI scripts), it provides database connectivity and there is even a bioperl project which provides many tools biologists need. This book can introduce you to only the very basics of `perl`. Still, you will get to the point where you learn how to write your own modules, generate small database files and perform dynamic programming.

1.7 Prerequisites

In order to perform the exercises you need to have access to a computer running either Linux, Unix or Mac OS X (the newest Apple operating system)

or the free Windows Unix emulator CygWin (see below). As you will learn soon, these systems are very similar. Thus, all the things we are going to learn will work on all Linux, Unix and Mac OS X computers. On a normal installation all required programs should be installed. Otherwise, contact the system administrator. My personal recommendation is to start with *Knoppix* Linux (see Sect. 2.9 on page 20). Knoppix runs from a CD-ROM and it requires no installation on the hard disk drive. Neither your operating system nor the data on your computer will be touched.
Alternatively, you can install the free *Cygwin* Unix emulator on a computer running the Microsoft Windows operating system (from Win95 upwards, excluding WinCE) (see Sect. 2.8.2 on page 19) or the commercial *VMware* package in order to run two operating systems on one computer (see Sect. 2.8.1 on page 19).

1.8 Conventions

What you see on your computer screen is written in `typewriter` style and boxed. I will refer to this as the *Terminal*. The data you have to enter are given behind the $ character. Key labels are written in a box. For example, the key labelled "Enter" would be written as [Enter]. Commands that appear in the text are written in `typewriter`, too. When necessary, space characters are symbolized by a "␣". Thus, "␣␣␣" means that you have to type three consecutive spaces. In the following example you would type `date` as input and get the current date as output.

1: Date

```
1  $ date
2  Thu Feb 13 18:53:26 CET 2003
3  $
```

In most cases you will find some text behind the terminal which describes the terminal content: in Terminal 1, line 1, we check for the current date and time.
Boxes labelled "Program" contain script files or programs. These have to be saved in a file as indicated in the first or second program line: `# save as hello.sh`. You will find the program under the same name on the accompanying website. As terminals, programs are numbered.

Program 1: Test

```
1  #!/bin/sh
2  # save as hello.sh
3  # This is a comment
4  echo "Hello World"
```

At the end of most chapters you will find exercises. These are numbered, too. The solution can be found in Chapter A.8 on page 273.

Part II

Computer and Operating Systems

2

Unix/Linux

This chapter's intention is to give you an idea about what operating systems in general, and Unix/Linux in particular, are. It is not absolutely necessary to know all this. However, since you are going to work with these operating systems you are taking part in their history! I feel that you should have heard of (and learned to appreciate) the outline of this history. Furthermore, it gives you some background on the system we are going to restrain; but if you are hungry for practice you might prefer to jump immediately to Chapter 3 on page 27.

2.1 What Is a Computer?

First things first! What is really going on when we switch on a computer? Well, if the power supply is plugged in, the main units will start to be alive. The first thing that starts is the BIOS (basic input/output system). The BIOS has all the information, e.g. which processor type is installed, what size the hard disk drive has, if there is a floppy disk or CD-ROM drive and so on, to initiate booting. The most important computer components are the *processor* (CPU, central processing unit), some kind of *memory* (RAM: random access memory; HDD: hard disk drive; FDD: floppy disk drive) and the *input-output devices* (usually a keyboard and the screen). Figure 2.1 on the following page shows a microprocessor system, which has all components necessary to be called a computer.

As important as the hardware (those things you can touch) is the software (application programs that you run, like a word processor) of a computer. The heart, that is the immediate interface between the hardware and the software, is the *operating system*. An operating system is a set of programs that control a computer. It controls both hardware and software. Most desktop computers have a single-user operating system (like Windows 2000 or lower, or Mac OS 9 or lower), which means that only one person can use the computer at

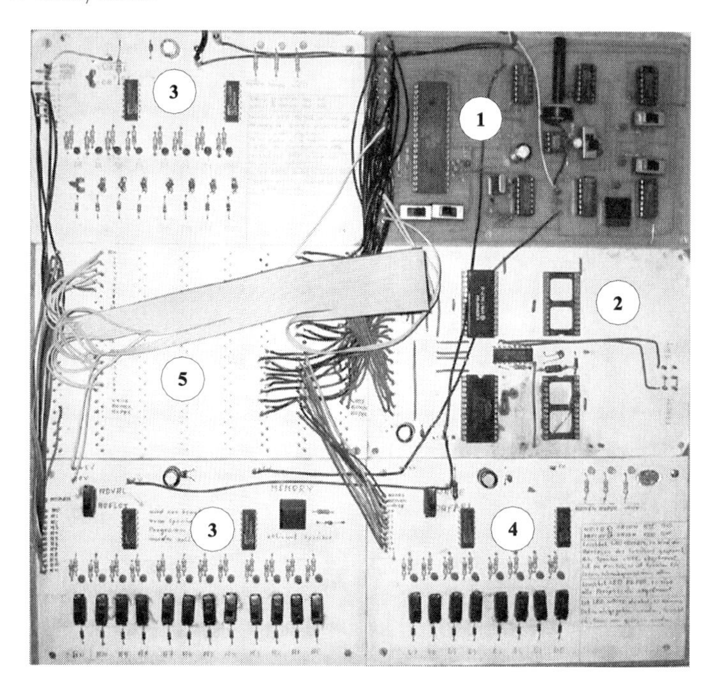

Fig. 2.1. A simple computer based on Zilog's Z80 processor (*1*). The system is equipped with 4 Kilobyte RAM (random access memory) (*2*). Data input and output are realized with electric switches and lamps, respectively (*3,4*). All components are connected via the system bus (*5*). The speed of the processor can be adjusted between single step and 1 MHz. The whole system was built by the author in the late 1980s

a time. Furthermore, many older operating systems, such as MS-DOS (MicroSoft Disk Operating System), can even handle only one job at a time; but most computers, especially in these days, are able to execute dozens of applications in parallel. In order to use the computer's processor efficiently, one should provide it with several tasks. This is what multitasking operating systems, such as Unix or Linux, do; but they go even a bit further and allow access for several users in parallel. That is called a multi-user operating system. Thus, with Unix or Linux many people can use the computer at the same time and several jobs can be run in parallel. Of course, every user has to login to his account on the computer from his own terminal, consisting of at least a keyboard and a screen connected to a very simple and cheap computer.

2.2 Some History

In the early 1960s, computers were as large as a wardrobe, expensive, had little memory and processor power and could only be used by single users.

In this environment, Ken Thompson and Dennis Ritchie, who worked for the Bell Laboratories of AT&T, developed the first version of Unix in 1969 (Fig. 2.2 on the next page). At this time, Unix was based on the research operating system *Multics* (multiplex information and computing system). Multics was an interactive operating system that was designed to run on the GE645 computer built by General Electric. With time, Multics developed into Unix, which ran on the PDP-7 computer built by Digital Equipment corporation (today Compaq). The new operating system had multitasking abilities and could be accessed by two persons. Some people called it Unics (uniplexed information and computer system). However, the operating system was limited to the PDP-7 hardware and required the PDP-7 assembler (this is the part which translates the operating system commands to processor code). It was Dennis Ritchie from Bell Laboratories who rewrote Unix in the programming language C. The advantage of C is that it runs on many different hardware systems – thus, Unix was now portable to hardware other than PDP-7 machines. Unix took off when Dennis Ritchie and Ken Thompson published a paper about Unix in July 1974 [10]. In the introduction to their paper they wrote: "Perhaps the most important achievement of UNIX is to demonstrate that a powerful operating system for interactive use need not be expensive either in equipment or in human effort: UNIX can run on hardware costing as little as $40,000, and less than two manyears were spent on the main system software."

2.2.1 Versions of Unix

With time, Unix became popular among AT&T companies, including Bell Laboratories and academic institutions like the University of California in Berkeley. From 1975 on, the source code of Unix was distributed for a small fee. From that point on, Unix started to take off (Fig. 2.2 on the following page).

Were there only nearly a dozen world-wide Unix installations in spring 1974, three dozen were registered in spring 1975, around 60 in autumn 1975 and 138 in autumn 1976. AT&T itself was not very interested in selling Unix. Thus, from 1975 on, Unix's development took largely place outside Bell Laboratories, especially at the University of California in Berkeley. In contrast to proprietary operating systems like Microsoft's DOS, Unix was now developed and sold by more than 100 companies including Sun Microsystems (Solaris), Hewlett Packard (HP/UX), IBM (AIX) and the Santa Cruz Operation (SCO-Unix). Of course, this led to the availability of different and partially incompatible Unix versions, which is a major problem for software developers. However, two Unix variants floated on top of all distribution:

AT&T System V – Developed in 1976 by the Bell Laboratories, the AT&T System V was distributed to many international universities. In 1989 AT&T founded the Unix-System Laboratories (USL) for further development of the

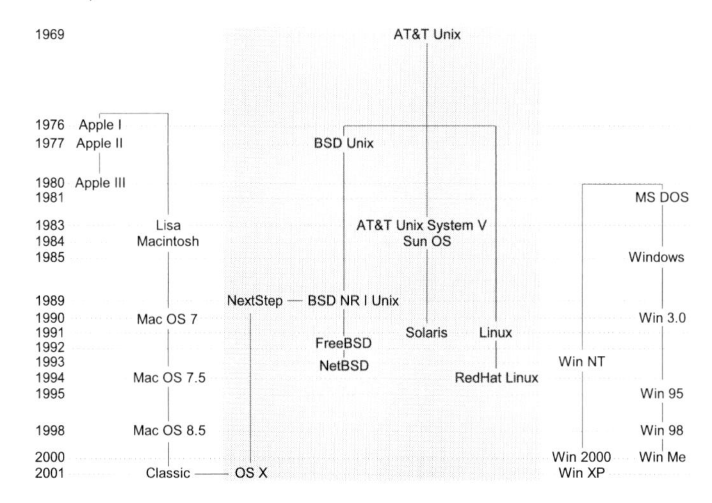

Fig. 2.2. Evolution of operating systems. Note the recent hybridization of Unix and Macintosh with Mac OS X in 2001. Unix-based operating systems are *underlaid in grey*

source code of System V. In 1991, a cooperation between USL and the company Novell started, which led to UnixWare running on Intel platforms. In 1996, Novell sold its Unix department to the Santa Cruz Operation (SCO), which is still developing Unix.

BSD v4 (Berkeley Software Distribution) – In parallel to AT&T, the University of California in Berkeley developed Unix up to version BSD v4. This version was running on Vax computers. In 1991, the spin-off company Berkeley Software Design Inc. was founded and, from that year on, has been selling BSD-Unix commercially. However, in parallel, there are still freeware versions available, namely FreeBSD and NetBSD. Apple's new operating system Mac OS X is actually based on BSD (see Sect. 2.7 on page 18).

Thus, the presently available Unix versions are based on either AT&T System V or BSD v4 (Fig. 2.2). In order to secure the compatibility between different Unix version *The Open Group* was founded in 1996.

2.2.2 The Rise of Linux

Linux is a very young operating system. Its first version was distributed by Linus Torvalds in 1991. The following news group posting was the first official announcement of the new operating system.

The rise of Linux

```
From: torvalds@klaava.Helsinki.FI (Linus Benedict Torvalds)
Newsgroups: comp.os.minix Subject: What would you like to see most
in minix? Summary: small poll for my new operating system
Message-ID: <1991Aug25.205708.9541@klaava.Helsinki.FI> Date: 25
Aug 91 20:57:08 GMT Organization: University of Helsinki Hello
everybody  out there using minix - I'm doing a (free) operating
system (just a hobby, won't be big and professional like gnu) for
386(486) AT clones. This has been brewing since april, and is
starting to get ready. I'd like any feedback on things people
like/dislike in minix, as my OS resembles it somewhat (same
physical layout of the file-system (due to practical reasons)
among other things). I've currently ported bash(1.08) and
gcc(1.40), and things seem to work. This implies that I'll get
something practical within a few months, and I'd like to know what
features most people would want. Any suggestions are welcome, but
I won't promise I'll implement them :-) Linus
(torvalds@kruuna.helsinki.fi) PS. Yes - it's free of any minix
code, and it has a multi- threaded fs. It is NOT protable (uses
386 task switching etc), and it probably never will support
anything other than AT-harddisks, as that's all I have :-(.
```

At that time the Finn Linus Torvalds was a student at the University of Helsinki (Finland). Since he offered the software freely and with source code via the internet, many programmers around the world had access to it and added more components to it, like improved file organization, drivers for different hardware components and tools like a DOS emulator. All these enhancements were again made available for free, including the source code. Thus, if there was any error with a software component it could be fixed by experienced programmers. A very important part of that concept was established already in 1984 by Richard Stallman with the *GNU* (GNU is not Unix) and *FSF* (Free Software Foundation) projects. GNU programs are freely available and distributable under the *General Public Licence* (GLP). In short, this means that GNU software can be used, developed and redistributed for free or commercially. However, the source code must always be a free part of the distribution. Many Unix users used GNU programs as substitute for expensive original versions. Popular examples are the text editor `emacs`, the GNU C-compiler (`gcc`) and diverse utilities like `grep`, `find` and `gawk`. Almost all open-source activities are now under the roof of the *Open Source Initiative* (OSI) directed by Eric S. Raymond.

As a matter of fact, Linux was not developed out of the blue but, from the beginning on, used GNU elements (like the operating system Mimix). Once the GNU C-compiler was running under Linux, all other GNU utilities could be compiled to run under Linux. Only the combination of the Linux kernel with GNU components, the network software from BSD-Unix, the free *X Windows*

System (see Sect. 2.3) from the MIT (Massachusetts Institute of Technology) and its *XFree86* port for Intel-powered PCs and many other programs, converted Linux to a powerful operating system which could compete with Unix.

2.2.3 Why a Penguin?

You might already have come across the nice penguin with his large orange feet and beak. The penguin is the mascot and symbol for Linux. Why is this? Well, although once bitten by a penguin in an Australian zoo, Linus Torvalds (the inventor of Linux) loves penguins [16]. Once the idea to use a penguin as logo for Linux was born, Linus Torvalds screened a variety of suggestions and finally chose the version from Larry Ewing, a graphics programmer and assistant system administrator at the Institute for Scientific Computing at the A&M University in Texas, USA. The penguins name is *Tux*, which is derived from the dinner jacket called tuxedo (short, tux).

2.2.4 Linux Distributions

There is one major problem when many people around the world develop and update an operating system: it is an awful task to collect all necessary and up-to-date components from the internet. Thus, different companies appeared, which took over the job and distributed complete sets of the Linux kernel with software packages, drivers, documentation and more or less comfortable installation programs and software package managers. Currently, there are many different Linux distributions on the market, the most famous among them are: Red Hat, SuSE, Debian, Mandrake and Knoppix. The good thing with Knoppix Linux is that you can run it from the CD-ROM. It does not require any installation and does not touch your hard disk drive at all, although you have access to it if you want. Therefore, it is well suited for playing around and I highly recommend its use for all the exercises in this book (see Sect. 2.9 on page 20).
Finally, there are some minimal distributions available on the net. This means you can have Linux on two floppy disks or so and install a minimum system on an old computer (i396 or i486).

2.3 X-Windows

We will work only with the command line. That looks a bit archaic, like in the good old DOS times. Anyway, Linux also provides a *graphical user interface* (GUI). In fact, it provides many GUIs and thus is much more powerful than Microsoft Windows. Graphical user interfaces are systems that make computing more ergonomic. Basic elements are windows which are placed on the

screen (desktop). Nowadays, no operating system without a graphical user interface could survive. I guess you know what I am talking about from your Windows system.
The GUI for Unix-based operating systems is called *X Windows System*, or just X. It was developed in 1984 at the Massachusetts Institute of Technology (MIT) and initially called "X Version 10". It was part of a larger project called Athena. The goal of that project was to integrate many different devices in one GUI. In 1987 "X Version 11, Release 1" was made public. From then on the software was developed by the community and is now available as "X Version 11, Release 6", also called "X11R6".
X-windows is completely network-based. This means that the output of a program on computer A can be visualized on a screen connected to computer B. Computers A and B can be at completely different places on earth. This has the great advantage that one can run programs on a powerful remote computer and visualize the output on a rather simple local machine.
Since there is a great diversity of graphical user interfaces, things are a bit different when you change from one system to another. In 1993, the companies developing and distributing Unix (Hewlett-Packard, IBM, The Santa Cruz Operation Inc., Sun-Soft, Univel and Unix Systems Laboratories) formed an alliance and agreed on a common standard called *Common Desktop Environment* (CDE). CDE is a commercial product based on X11R6. The most common desktop environments for Linux are *KDE* (*www.kde.org*) and *Gnome* (*www.gnome.com*). Both are freely available.

2.3.1 How Does It Work?

The graphical desktop can be split into several parts: the *X-server*, the *X-clients*, the *window manager* and the *desktop*; but the Unix/Linux desktop is something different from a Windows desktop.
The X-server is hosted by the computer (or graphical terminal station) you are sitting in front of. An X-server is nothing more than the black and white chess pattern you see very shortly before the graphical background of your window manager appears. The X-server handles the graphical presentation and is responsible for the communication between the hardware (in particular the graphic card, which handles the screen) and the software (the X-programs). If you would run only an X-server, no graphics would be possible. It offers neither a menu, nor windows or any other features you need. Here the X-client comes into play. The work of the X-server is done with X-clients. These X-clients use libraries that are integrated in the X-server and contain the information how to display graphics. If you see an X-terminal (console) on your monitor, it is an X-client. The communication between X-server and X-client works through the network. That is the reason why you can start an X-client on any computer in the network and see it somewhere else (this is the way the X-terminals work; you start the programs on a powerful server and sit in front of a simple terminal). Even if you work on an

isolated computer and have no network card installed, X-client and X-server communicate via the network. In that case, Unix/Linux simulates a network (loopback). X-server and X-clients alone are not really comfortable to work with. Useful functions like "Maximizing", "Minimizing" and "Close Window" are not included in the functionality of the X-server and X-clients but made available by a window manager. There are several window managers available, like FVWM (*www.fvwm.org*), IceWM (*www.icewm.org*), Window Maker (*www.windowmaker.org*), Sawfish, formerly Sawmill (*sawmill.sourceforge.net*) and Metacity (*www.gnome.org/ softwaremap/projects/Metacity*).
A long time ago, only these three components (X-server, X-client and window manager) existed. However, in the past years, an additional "thing" has come into existence: the desktop. The desktop offers functionalities similar to those we are used to from Windows, like putting program and file icons onto the desktop, which can be started or opened by double-clicking, respectively. However, with Unix/Linux, usually a single click is sufficient. Of course, icons existed long before (e.g. after minimizing a window with FVWM), but the functionality was not as great as with a desktop. KDE and Gnome are two popular desktops running on Unix/Linux. They come with their own window managers, but both programs could be used with alternative window managers as well.

2.4 Unix/Linux Architecture

We have already heard that Unix and Linux are operating systems. How are they organized? In principle, we distinguish four components: the *hardware*, the *kernel*, the *shell* and *applications* (Fig. 2.3).

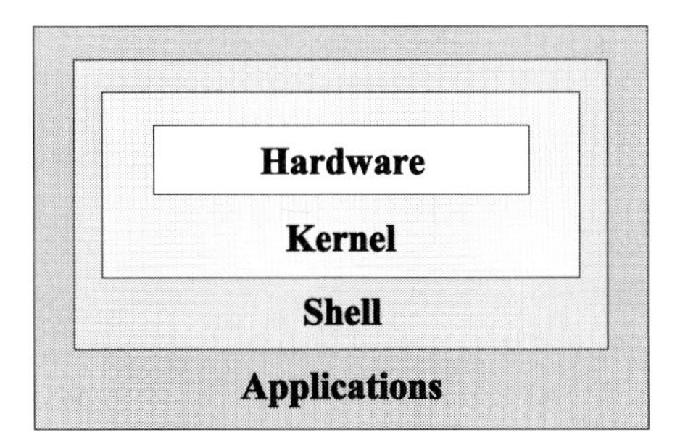

Fig. 2.3. The kernel is the heart of Unix/Linux. It is wrapped around the hardware and accessed via the shell

The hardware, i.e. the memory, disk drives, the screen and keyboard and so on, is controlled by the kernel. The kernel is, as the name implies, the heart of Unix/Linux. Only the kernel has direct control over the hardware. Unix/Linux's kernel is programmed in C. It is continuously improved; this means errors are corrected and drivers for new hardware components that are available are included. Thus, when you buy a Unix/Linux version, the kernel version is usually indicated. The newest kernel is not necessarily the best: it might have new errors. For us, the users, it is very uncomfortable to communicate with the kernel. Therefore, the shell was developed. The shell is an envelope around the kernel. It provides commands to work with files or access floppy disk drives and such things. You are going to learn more about the shell in Chapter 7 on page 81. Finally, there are the applications like OpenOffice or ClustalX or `awk` or `perl`. These communicate with the kernel, usually via the shell. You see, the Unix/Linux operating system is clearly structured. This is part of Unix's philosophy: "Keep it simple, general and extensible".

2.5 What Is the Difference Between Unix and Linux?

Essentially: the price! And: Linux is not free Unix but a Unix-like operating system! Otherwise, for the user, the differences are almost invisible. One point is that Linux runs on a wide range of different processors. You probably have heard of Intel processors like Pentium, Celeron and so on. Most personal computers have processors produced by either Intel or AMD, or, in the case of Apple computers, PowerPC processors produced by Motorola and IBM; but there are many more processors on the market like SPARC from Sun Microsystems or MIPS and Alpha from DEC (Digital Equipment Corporation). Linux has been ported (adapted) to all these processors. Unix is much more restricted in this respect.

2.6 What Is the Difference Between Unix/Linux and Windows?

Essentially: compatibility with fancy-dancy external devices! It should be clearly stated that Windows is usually easier to use than Unix/Linux. Although Unix/Linux are improving and in these days almost keep up with Windows, they still fall behind when it comes to comfort. The reason is that Windows is much more widespread and almost all companies design their software for the Windows platform. One tends to ask: why is that? Well, if, e.g. Windows 2000 is up and running on a computer, then this forms a clearly defined environment. This is also true for kernels; however, they develop much faster and exist in many more versions.
Of course, apart from these ergonomic differences there are some hard facts, which make both operating systems clearly different. For example, with the exception of Windows NT, which also runs on Alpha processors, the desktop

and server-oriented Windows operating systems run only on Intel and AMD processors. Furthermore, Unix-based operating systems are much closer to the data. Thus it is much easier to format and analyze files.

2.7 What Is the Difference Between Unix/Linux and Mac OS X?

Essentially: None! In fact, the latest version of Apple's operating system (*Mac OS X*, Macintosh operating system 10) is based on BSD, which is a Unix clone (see Sect. 2.2.1 on page 11). If you want to work on a Unix-like terminal, you can run the application *Terminal*. It is a regular program that can be found under "Applications → Utilities". It is possible to do everything you learn in this book in the Mac OS X Terminal, too. If you encounter problems, please report them to me. Apple has also developed its own graphical user interface, called *Aqua*. It basically works as X-Windows (see Sect. 2.3 on page 14) does. If you wish, you can even install native X-Windows and run it on the Mac. I must admit that Aqua is the crown jewel of Mac OS X. It is light years ahead of KDE or Gnome (see Sect. 2.3 on page 14) and makes Mac OS X the most user-friendly Unix-based operating system around. There has been a whole bunch of software ported to Mac OS X, including *OpenOffice* and the graphic program *Gimp*. There is an interesting project called *Fink*. The Fink Project wants to bring the full world of Unix open-source software to Mac OS X. They modify Unix software so that it runs on Mac OS X and make it available for download as a coherent distribution. The Fink Project uses a special tool in order to facilitate easy installation procedures. Take a look at *fink.sourceforge.net*.
Why did Apple switch to Unix? In 1985, Steve Jobs, the founder of Apple, left to start the company NeXT, whose NeXTStep operating system was based on BSD Unix. When Apple bought NeXT in 1996, Jobs, NeXTStep (then called OpenStep) and its Terminal program came along with it. From there, it was only a small step to fuse both systems.

2.8 One Computer, Two Operating Systems

You might not wish to eliminate your Windows operating system before knowing how much you like Unix or one of its derivatives. A common solution to this problem is to install both operating systems on the same computer as a *dual boot* system. This is only possible with Linux and Windows, since both can use the same processor architecture (see Sects. 2.5 on the preceding page and 2.6 on the page before). When you start up you have to choose which operating system you want to work with. If you want to write a document in *Microsoft Word*, you can boot up in Windows; if you want to run *Genesis*, a general-purpose neural network simulator for Linux, you must shut down your Windows session and reboot into Linux. The problem is that you cannot do both at the same time. Each time you switch back and forth between Windows and Linux, you have to reboot again. This can quickly get tiresome. Therefore, you might want to use two operating systems on one computer – in parallel! There are several possibilities.

2.8.1 VMware

A powerful solution is offered by the software package *VWware*. With VMware you can either run Linux on top of Windows or vice versa. The great advantage lies in the fact that you really run both operating systems at the same time. You switch to the other operating system as you would switch from one program to another – just activate the application window. Furthermore, you run the real operating system, no emulator, which only imitates an operating system as is the case with CygWin (see below). The disadvantages, namely calculation power and high RAM requirements, are out of date. Modern supermarket computers for 400 Euro or US$ already offer enough performance. Take a look at their homepage (*www.vmware.com*) for more information.

2.8.2 CygWin

Cygwin is a Unix environment for Windows. It consists of two parts: a) a dynamic link library (cygwin1.dll), which acts as a Unix emulation layer providing substantial Unix API (application programming interface) functionality and b) a collection of tools, ported from Unix, which provide Unix/Linux look and feel. In addition, CygWin is really easy to install. Take a look at *cygwin.com* about the installation procedure and make sure you install the following packages: `bash`, `gawk`, `sed`, `grep` and `perl`.

2.8.3 Wine

Wine is a Unix implementation of the Win32 Windows libraries, written from scratch by hundreds of volunteer developers and released under an Open Source license. Thus, Wine is the counterpart of CygWin. The Wine project (*www.winehq.com*) started in 1993 as a way to support running Windows 3.1 programs on Linux. Bob Amstadt was the original coordinator, but turned it over fairly early on to Alexandre Julliard, who has run it ever since. Over the years, ports for other Unix versions have been added, along with support for Win32. Wine is still under development, and it is not yet suitable for general use. Nevertheless, many people find it useful in order to run a growing number of Windows programs. There is an application database for success and failure reports for hundreds of Windows programs.

2.8.4 Others

Of course, there are many other possibilities and you should check out yourself which solution is suitable for your setup. With *Win4Lin* you can run Windows programs under Linux (*www.netraverse.com*). Another option might be *Plex86* (*plex86.sourceforge.net*). Plex86 is an extensible free PC virtualization software program which will allow PC and workstation users to run multiple operating systems concurrently on the same machine. Plex86 is able to run several operating systems, including MSDOS, FreeDOS, Windows9x/NT, Linux, FreeBSD and NetBSD. It will run as much of the operating system and application software natively as possible, the rest being emulated by a PC virtualization monitor. I am sure there is more software available and even more to come. Check out yourself.

2.9 Knoppix

Knoppix is a free Linux distribution developed by Klaus Knopper, which can be run from a CD-ROM. This means you put the CD-ROM into your computer's CD-ROM drive, switch on the computer and start working with Linux. Knoppix has a collection of GNU/Linux software, automatic hardware detection and support for many graphics cards, sound cards, SCSI and USB devices and other peripherals. It is not necessary to install anything on the hard disk drive. Due to on-the-fly decompression, the CD carries almost 2 GB of executable software. You can download or order Knoppix from (*www.knopper.net/knoppix*).

The minimum system requirements are an Intel-compatible CPU (486, Pentium or later), 20 MB of RAM for the text-only mode or at least 96 MB for the graphical mode with KDE. At least 128 MB of RAM is recommended if you want to use OpenOffice. Knoppix can also use memory from the hard disk drive to substitute missing RAM. However, I cannot recommend using this option, since it decreases the performance. Take a look at the Knoppix manual in order to learn how to do this. Of course, you need to have a bootable CD-ROM drive or a boot floppy if you have a standard CD-ROM drive (IDE/ATAPI or SCSI). For the monitor, any standard SVGA-compatible graphic card will do. If you are one of those freaks playing the newest computer games and always buying the newest graphic card you might run into trouble; but I am sure that you then own at least one out-dated computer you can use for Linux. As a mouse you can either use a standard serial or PS/2 or an IMPS/2-compatible USB-mouse.

Before you can start Knoppix, you need to change your computers BIOS (basic input/output system) settings to boot from the CD. When you start up your computer you are normally asked whether you want to enter the system setup (BIOS). Usually one of the following key (combinations) is required: (Del) or (Esc) or (F2) or (Ctrl)+(Alt)+(Esc) or (Ctrl)+(Alt)+(S). When you succeed in changing the settings you are done; but be careful not to change anything else! That might destroy your computer. However, if your computer does not support the option to boot from the CD-ROM, or you are afraid of doing something wrong, you have to use a boot disk. You can create this disk from the image in the file */KNOPPIX/boot.img* on the CD-ROM. Read the manual on the Knoppix CD-ROM for more details on this issue. Once prepared, you put the CD in the drive and power up the computer. After some messages the system halts and you see the input prompt "`boot:`". Now you can hit (F2) and optimize Knoppix to your needs. For example, `knoppix lang=de` enables the German keyboard layout.

2.10 Software Running Under Linux

Of course, it is not my intention to name all available software packages here. The objective is to give you an overview of which software is available and which Windows-based software it substitutes. Needless to say that almost all software is freely available.

2.10.1 Biosciences

There is a lot of software available for all fields of academics. Being a biologist myself, I will restrict myself to list some important biological software packages available for Linux. An updated list can be obtained from *bioinformatics.org/software*.

Emboss – Emboss is the **E**uropean **M**olecular **B**iology **O**pen **S**oftware **S**uite. Emboss is freely available and specially developed for the needs of molecular biologists. Currently, Emboss provides a comprehensive set of over 100 sequence analysis programs. The whole range from DNA and protein sequence editing, analysis and visualization is covered. Restriction analysis, primer design and phylogenetic analysis can all be performed with this software package. Take a look at *www.hgmp.mrc.ac.uk/Software/EMBOSS*.

Staden Package – The Staden Package is a software package free for academics (charge for commercial users) including sequence assembly, trace viewing/editing and sequence analysis tools. It also includes a graphical user interface to the Emboss suite. More information can be obtained at *www.mrc-lmb.cam.ac.uk/pubseq/staden_home.html*

Blast – Blast (**B**asic **L**ocal **A**lignment **S**earch **T**ool) is a set of similarity search programs designed to explore online sequence databases. Alternatively, you can set up your own local databases and query those. In Chapter 5 on page 53 we will download, install and run this program.

ClustalW – ClustalW is probably the most famous multiple-sequence alignment program available. It is really powerful and can be fine-tuned by a number of program options. In Chapter 5 on page 53 we will download, install and run this program.

Dino – Dino (*www.dino3d.org*) is a real-time three-dimensional visualization program for structural biology data. Structural biology is a multidisciplinary research area, including X-ray crystallography, structural NMR, electron microscopy, atomic-force microscopy and bioinformatics (molecular dynamics, structure predictions, surface calculations etc.). The data produced by these different research areas are very diverse: atomic coordinates (models and predictions), electron density maps, surface topographs, trajectories, molecular surfaces, electrostatic potentials, sequence alignments etc... Dino aims to visualize all these structural data in a single program and to allow the user to explore relationships between the data. There are five data types supported: structure (atomic coordinates and trajectories), surface (molecular surfaces), scalar fields (electron densities and electrostatic potentials), topographs (surface topography scans) and geoms (geometric primitives such as lines). The number and size of the data the program can handle is limited only by the amount of working memory (RAM) present in the system.

2.10.2 Office and Co.

Apart from a growing number of science-related programs, Unix/Linux gets more and more attractive for personal users. This is because almost everything known from Windows is nowadays also available for Unix/Linux – and more...

Graphical Desktop – All versions of Microsoft Windows desktops look pretty much the same. This is an advantage when it comes to, e.g. changing from one version to another. However, it can be boring. It does not matter if you install different themes because it would function in the same way as before, and you cannot change the functionality. If you are missing a function you can only hope that it will become available later – when you have to buy it, of course. Unix/Linux has a completely different philosophy about graphical desktops. Variety is the theme (see Sect. 2.3.1 on page 15)!

Editors – You can choose between a variety of different text editors. There is the whole range from single-line to graphical editors available. Just to name a few: `emacs`, `vi` (see Sect. 6.3 on page 72), `ed`, `pico` (see Sect. 6.2 on page 70), `gedit`, `kedit`.

Office Applications – With *OpenOffice* (*www.openoffice.org*) there is a whole bunch of office applications available. Among these are Writer (like MS Word), Calc (like MS Excel), Impress (like MS Powerpoint), Draw (like Corel Draw), HTML Editor, Math Editor and others. Common MS Office files can be opened, edited and even saved in MS Office format. OpenOffice uses for its files the markup language XML (Extended Markup Language). Furthermore, the files are automatically compressed.
Of course, there are more options. *KOffice*, integrated to the KDE desktop, offers office applications and there are a number of stand-alone programs.

Image Processing – *Gimp* (*www.gimp.org*) is a very powerful image processing with a number of plugins and which can do basically everything Adobe Photoshop can do.

Vector Graphics – Apart from OpenOffice Draw you can work, for example, with *XFig* (*www.xfig.org*) to create vector graphics.

Others – Okay, as you can imagine there are zillions of software packages available. Basically, all needs will be satisfied. You can synchronize your Palm, install relational databases, choose from a bunch of web browsers and email programs and so on...

Part III

Working with Unix/Linux

3

The First Touch

In this chapter you will learn the basics in order to work on a Unix-based computer: *login*, execute *commands*, *logout*. Everything you are going to do is happening at the command line level. This means, the look and feel will be like in good old DOS times. It will look like stone-age computing. However, you should remember that although a graphical interface is often nice and comfortable, it consumes a lot of power and only hinders us from learning what is really important. Furthermore, the Unix/Linux command line is extremely powerful. You will soon get accustomed to it and never want to miss it again. Let us face it...

3.1 Login

The process of making yourself known to the computer system and getting to your Unix/Linux account is called logging in. There are two prerequisites: first you need to connect to the computer, and second you must have an account on that computer. That is like withdrawing money from a bank account. You must identify yourself to the cash machine before you get any money and you must, of course, have a bank account (with some money in it). There are several ways to connect to a Unix-based computer.

3.1.1 Working with CygWin or Mac OS X

In case you want to work with Apple's Mac OS X operating system (see Sect. 2.7 on page 18) or the Unix emulator CygWin, which runs in the Windows environment (see Sect. 2.8.2 on page 19), then you need not care about any password. When you start CygWin, you will directly end up at the command line of the bash shell. In Mac OS X you find the *Terminal* application under "Applications → Utilities". Again, starting the application will immediately bring you to the command line. In any case you should look up your username with the command `whoami` (just type `whoami` and then press (Enter)).

3.1.2 Working Directly on a Unix/Linux Computer

You might have installed Linux on your computer or run it from a CD-ROM on your computer (like the Knoppix distribution of Linux, see Sect. 2.9 on page 20). In that case, you boot the computer and get directly to the login screen. Depending on the system settings, this might be either a graphical login or a command line login. In both cases you enter your *username* and *password* (with Knoppix you do not need a username and password; you are directly locked in as user *knoppix*). With the graphical login you immediately end up on a nice-looking graphical desktop – most probably KDE or Gnome. In that case, you have to open a terminal window in order to follow the examples given in this script. If the command line login appears after booting the computer you have to follow the instruction given in Terminal 2.

3.1.3 Working on a Remote Unix/Linux Computer

A quite common tool to connect to a remote computer is *telnet*. However, it is very insecure. Nowadays, most Unix/Linux systems do not allow telnet sessions. The freeware Windows application PuTTY allows one to connect to a Linux Server with a secure protocol (SSH). The program can be downloaded at: *www.chiark.greenend.org.uk/~sgtatham/putty*
Although there are many possible settings, you basically need the IP address of the computer on which you have an account (see Fig. 3.1 on the next page).

After connecting, a login window pops up. Here you have to enter your username and password. The windows content might look pretty much like Terminal 2.

Terminal 2: Login

```
1  login as: Freddy
2  Freddy@134.95.189.1's password:
3  Last login: Mon Mar 17 14:04:07 2003 from 134.95.189.37
4  [Freddy@nukleus Freddy]$
```

In Terminal 2, line 1, the username `Freddy` is entered. Then line 2 pops up asking for the password of Freddy at the computer named `134.95.189.1`. Note: Usually the computer's name is not a number but a name. You will not see anything happening on the screen while you type in your password. You type blind. This is for reasons of security. After successful login, lines 3 and 4 pop up. Line 3 shows the date of your last login and from where you last logged in (in this case from the computer with the network identification number 134.95.189.37). Finally, as shown in line 4, you are logged in to the system and the "`$`" character indicates that the command line is ready for input. Line 4 is also called the *shell prompt*, the dollar character "`$`" is the prompt character. Shell prompts usually end with $ or %. The prompt can be customized, thus your own shell prompt might look different. A prompt that ends with the hash character (#) usually means that you are logged in

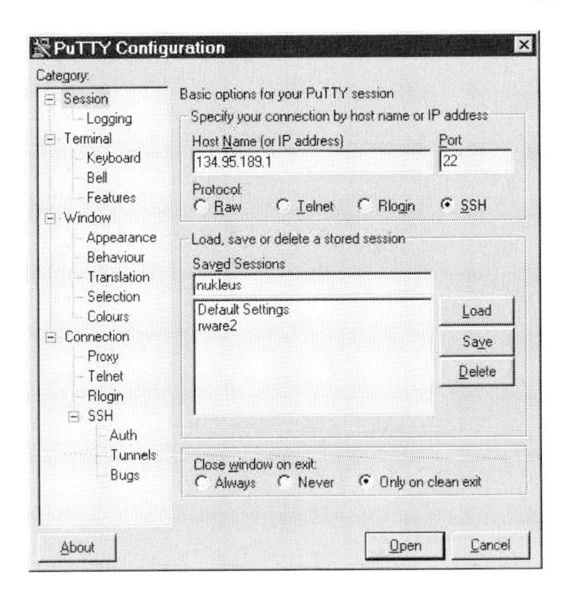

Fig. 3.1. The PuTTY configuration window. The most important information needed is the IP address or host name of the computer you want to work on remotely

as *superuser* or *root*. The root account belongs to the administrator who has access to all system components and owns all rights. In Terminal 2 on the facing page the shell prompt tells you that you are connected as user Freddy to the computer nukleus and you are currently in the directory *Freddy* (that is Freddy's home directory). The messages you see at login differ from system to system; but this general outline should work out in most cases.
I guess it is clear that you have to press the (Enter) or (Return) key after you have entered your name and password, respectively. Furthermore, make sure to type your username at the *login* prompt and your password at the *password* prompt without errors. Backspacing to delete previous characters may not work (though Linux is more forgiving than Unix). If you make a mistake, repeatedly use the (Enter) key to get a new *login* prompt and try again. Also make sure to use the exact combination of upper- and lowercase letters. *Linux is case-sensitive!*
When you login to your account it might look different. However, as a general rule, you get three messages after a command line login: a) *motd* – This is the *message of the day*. Here the system administrator might inform you about system-down times or just present a message to cheer you up. b) *last login* – As mentioned above, this is the date of your last login. c) *You have new mail* – This message does exactly what it says. It tells you that at least one new email is waiting for you in your email program.

3.2 Using Commands

Once you have logged in to the computer you can start working. Let us start with some simple commands in order to get accustomed to the Unix/Linux environment. Commands are entered in the same way as you just entered your username: enter the command `date` and press the (Enter) key.

Terminal 3: Date

```
1  $ date
2  Tue Jan 23 17:23:43 CET 2003
3  $
```

As shown in Terminal 3, this command will show the computer system's date and will end with giving you an empty command line again. In fact, `date` is a program. It collects data from the system and uses the *standard output* to display the result. The standard output is the screen. You could also *redirect* the output into a file. This is illustrated in Terminal 4.

Terminal 4: Date to File

```
1  $ date > file-with-date
2  $
```

The character ">" redirects the output of the command `date` into a file which is named *file-with-date*. If it does not exist, this file is created automatically. Otherwise it will be overwritten. You could use ">>" to append the output to an existing file. Of course, since the output is redirected, you do not see the result. However, as shown in Terminal 5, you can use the command `cat` to display the content of a file.

Terminal 5: Show Date-File

```
1  $ cat file-with-date
2  Tue Jan 23 17:25:44 CET 2003
3  $
```

Most commands accept options. These options influence the behaviour of the command (which is, as said, in fact a program). Options are often single letters prefixed with a dash (-, also called hyphen or minus) and set off by spaces or tabs. For a list of all strange characters and their names see Section A.8 on page 272.

Terminal 6: Show Date-File

```
1  $ cat -n file-with-date
2       1  Tue Jan 23 17:25:44 CET 2003
3  $
```

In the example in Terminal 6 the option `-n` to the command `cat` prints line numbers.

3.2.1 Syntax of Commands

In general, commands can be simple one-word entries such as the `date` command. They can also be more complex. Most commands accept *arguments*. An argument can be either an *option* or a *filename*. The general format for commands is: `command option(s) filename(s)`. Commands are entered in lowercase. Options often are single characters prefixed with a dash (see Terminal 6 on the facing page). Multiple options can be written individually (`-a -b`) or, sometimes, combined (`-ab`) – you have to try it out. Some commands have options made from complete words or phrases. They start with two dashes, like `--confirm-delete`. Options are given before filenames. Command, option and filename must be separated by spaces. In a few cases, an option has another argument associated with it. The `sort` command is one example (see Sect. 6.1.1 on page 65). This command does sort the lines of one or more text files according to the alphabet. You can tell `sort` to write the sorted text to a file, which name is given after the option `-o` (output).

Terminal 7: Options

```
1  $ date > file-with-date
2  $ cat file-with-date
3  Tue Jan 23 17:25:44 CET 2003
4  $ date >> file-with-date
5  $ cat file-with-date
6  Tue Jan 23 17:25:44 CET 2003
7  Tue Jan 23 17:25:59 CET 2003
8  $ sort file-with-date
9  Tue Jan 23 17:25:44 CET 2003
10 Tue Jan 23 17:25:59 CET 2003
11 $ sort -o sorted-file file-with-date
12 $ cat sorted-file
13 Tue Jan 23 17:25:44 CET 2003
14 Tue Jan 23 17:25:59 CET 2003
15 $
```

Terminal 7 summarizes what we have learned up to now. You should now have a feeling for options. In line 1 we save the current date and time in a file named *file-with-date* (when you do the above exercises you will in fact overwrite the old file content). In line 2 we use the command `cat` to print out the content of *file-with-date* on the screen. With the next command in line 4 we add one more line to the file *file-with-date*. Note that we use `>>` in order to append and do not use `>` to overwrite the file. After applying `cat` again, we see the result in lines 6 and 7. We then sort the file with the command `sort`. Of course, that does not make sense, since the file is already sorted. The result is immediately printed onto the screen, except that we advise `sort` to do it differently. We do that in line 11. With the option `-o` (output) the result is saved in the file *sorted-file*. The file we want to sort is given as the last parameter. The result is printed in lines 13 and 14.

3.2.2 Editing the Command Line

What do we do if we have made a mistake in the command line? There are a number of commands to edit the command line. The most important ones are listed below:

BkSp	deletes last character
Del	deletes last character
Ctrl+H	deletes last character
Ctrl+U	deletes the whole input line
Ctrl+S	pauses the output from a program
Ctrl+Q	resumes the output after Ctrl+S

It is highly recommended not to use Ctrl+S. However, when you get experienced it might become useful for you.

3.2.3 Change Password

Together with your username your password clearly identifies you. If anyone knows both your username and password, he or she can do everything you can do with your account. You do not want that! Therefore you must keep your password secret. When you logged in for the first time to Unix/Linux you probably got a password from the system administrator; but probably you want to change it to a password you can remember. To do so you use the command `passwd` as shown in Terminal 8.

Terminal 8: passwd

```
1 $ passwd
2 Changing password for user Freddy.
3 Changing password for Freddy
4 (current) UNIX password:
5 New password:
6 Retype new password:
7 passwd: all authentication tokens updated successfully.
8 $
```

First you have to enter your current password. Then, in line 5 in Terminal 8 you are requested to enter your new password. If you choose a very simple password, the system might reject it and ask for a more complicated one. Be aware that passwords are case-sensitive and should not contain any blanks (spaces)! You should also avoid using special characters. If you log in from a remote terminal the keyboard might be different. To avoid searching for the right characters, just omit them right from the beginning. After you have retyped the password without mistake, the system tells you that your new password is active. Do not forget it!

3.2.4 How to Find out More

Imagine you remember a command name but you cannot recall its function or syntax. What can you do? Well, of course, you always have the option of trial and error. However, that might be dangerous and could corrupt your data. There are much more convenient ways.

First, many commands provide the option `-h` or `--help`. Guess what: *h* stands for help. Actually, `--help` works more often than `-h` and might be easier to remember, too. The command `ls -h`, for example, lists file sizes in a human readable way – you must use `ls --help` instead in order to obtain help. By applying the help option, many commands give a short description of how to use them.

There is, however, a much more informative way than using the help option. Almost all Unix systems, including Linux, have a documentation system derived from a manual originally called the *Unix Programmer's Manual*. This manual has numbered sections. Each section is a collection of manual pages, also called *manpages*, and each program has its own manpage. Most Unix/Linux installations have individual manpages stored on the computer. Thus they can be accessed at anytime. To do so, use the command `man` (manual). For example, if you want to find information about the command `sort`, then you type "`man sort`". Usually the output is directly sent to a page viewer (pager) like *less* (see Sect. 6.1.3 on page 67). If not, you can pipe it to `less` by typing "`man sort|less`". You can then scroll through the text using the (↑), (↓), (PgUp) and (PgDn) keys. You leave the manpage by pressing (Q). You should note that manpages are not available for all entries. Especially commands like `cd`, that are not separate Linux programs but part of the shell, are not individually documented. However, you can find them in the shell's manpages with `man bash`.

The `info` *`command`* (information) command serves a similar purpose. However, the output format on the screen is different. To read through the output press the space bar (Space); to quit the documentation press (Q). It is not possible to scroll through the output.

If you have a faint idea of a command use `apropos` *`keyword`*. This command searches for the *keyword* in an index database of the manpages. The keyword may also contain wildcards (see Sect. 7.10 on page 91).

3.3 Logout

Almost as important as to know how to login to a system, you should know how to logout. You cannot simply switch off the computer. Usually, the computer will be a big server sitting in another building or even city; but even if you run Linux at home on your desktop computer you should never, ever simply switch off the power supply. In case you work on a remote server you just have to logout. Several commands allow you to logout from the system.

The two most common ones are the command `logout` or the key combination Ctrl+D. You are done. If you work on your home computer you will see a new "login:" prompt. You should login again as superuser (root) and halt the system with the command `halt`. A lot of information will flush over the screen until it says something like: "Run Level 0 has been reached". Then you can switch off the power, if it does not do so automatically.

Exercises

There is no other way to learn Linux than by working with it. Therefore you must exercise and play around.

3.1. Login to your Unix-based computer. What information do you get from your computer?

3.2. Write a command and use the erase key to delete characters. The erase key differs from system to system and account to account. Try out BkSp, Del and Ctrl+H.

3.3. Use the command `date` and redirect the output into a file named *the_date*. Append the current date again to that file and print the resulting file content onto the screen.

3.4. Change your password. After you have changed it successfully, you might want to restore your old one. Do so!

3.5. Logout from your current session.

4

Working with Files

Before people started working with computers they used to work at desks (nowadays, people work with a computer on their lap on the sofa.) They had information in files which were organized in different folders. Well, now we work on virtual desktops and still use files and folders. Figure 4.1 gives an example of how files are organized in directories.

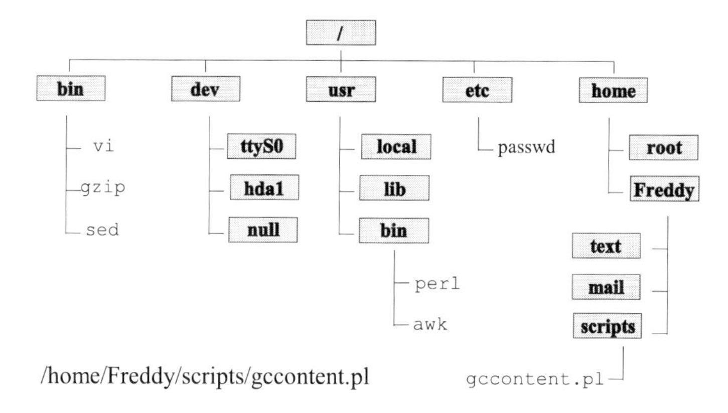

Fig. 4.1. This sketch illustrates the organization of directories (folders) and files in Unix/Linux. The root directory is shown *at the top*. Directories are *underlaid in grey*. As an example, the path to the file *gccontent.pl* is shown

Files are the smallest unit of information the normal user will encounter. A file might contain everything from nothing to programs to file archives. In order to keep things sorted, files are organized in folders (directories). You can name files and directories as you wish – but you must avoid the following special characters:

```
/ | \ - < > , # . ~ ! $ & ( ) [ ] { } ´ ‘ " ? ^
```

You also should avoid to use the space character "␣". In order to separate words in filenames the underscore character "_" is commonly used. Unix/Linux discriminate between upper- and lowercase characters. A file named *file* can be distinguished from a file called *File*. This is a very important difference to Microsoft Windows.
In fact, in Unix/Linux everything is a file. Even devices like printers or the screen are treated as if they were files. For example, on many systems the file *tty1* stands for the serial port number 1. Table 4.1 gives you a short description of the most common system directories and their content.

Table 4.1. Parameter that can be used with the command `chmod`

Directory	Content
/bin	essential system programs
/boot	kernel and boot files
/dev	devices like printers and USB or serial ports
/etc	configuration files
/home	user's home directories
/lib	important system files
/lost&found	loose file fragments from system checks
/mnt	external drives like floppy disk
/proc	system information
/root	root's home directory
/sbin	administrative programs
/tmp	temporary files
/usr	static files like programs
/var	variable files like log files

4.1 Browsing Files and Directories

A special case of files are folders or directories. They give the operating system a hierarchical order. In Unix/Linux, the lowest level, the root directory, is called "/". When you login to the computer you end up in your home directory which is "/home/username". In order to find out in which directory you are type `pwd` (print working directory).

Terminal 9: pwd

```
1 $ pwd
2 /home/Freddy
3 $
```

As shown in Terminal 9, the command `pwd` (print working directory) prints out the *path* to the home directory of the user Freddy. The path is specific for each file and directory and contains all parent directories up to the root directory "/". Thus, line 2 of Terminal 9 on the facing page reads: the directory *Freddy* resides in the directory *home*, which in turn resides in the directory *root* (/). The name of the home directory always equals the username, here *Freddy*. If you want to see what is in your home directory, type the command `ls` (list).

Terminal 10: ls

```
1  $ mkdir Awk_Programs
2  $ mkdir Text
3  $ mkdir .hidden
4  $ date > the.date
5  $ date > Text/the.date
6  $ ls Awk_Programs
7  Text  the.date
8  $ ls -a
9  .  ..  Awk_Programs  .hidden  Text  the.date
10 $ ls -l
11 total 12
12 drwxrwxr-x  2  rw  rw  4096  Apr 1 16:50  Awk_Programs
13 drwxrwxr-x  2  rw  rw  4096  Apr 1 16:51  Text
14 -rw-rw-r--  1  rw  rw    30  Apr 1 17:16  the.date
15 $ ls -lh
16 total 12K
17 drwxrwxr-x  2  rw  rw  4.0K  Apr 1 16:50  Awk_Programs
18 drwxrwxr-x  2  rw  rw  4.0K  Apr 1 16:51  Text
19 -rw-rw-r--  1  rw  rw    30  Apr 1 17:16  the.date
20 $ ls Text
21 the.date
22 $
```

In Terminal 10 we see the effect of different commands. In lines 1 to 3 we create the directories *Awk_Programs*, *Text* and *.hidden*, respectively. That is done with the command `mkdir` (make directory). Note that these directories are created in the current directory. In line 4 we create the file *the.date* with the current date (see Terminal 4 on page 30) in the current directory. In line 5 we create a file with the same name and content; however, it will be created in the subdirectory *Text* (a subdirectory is the child of directory; the root directory is the mother of all directories). Next, we apply the command `ls` to check what is in the current directory. This command prints out all directories and files in the current directory. Note that the directory *.hidden* is missing in the list. Since it is preceded by a dot the folder is hidden. – Note: Hidden files start with a dot. – By default the `ls` command does not show hidden files and directories. Actually, many system files are hidden. It prevents the directories from looking chaotic. In line 8 we add the option `-a` (all) to the `ls` command. With this option we force the list command to show hidden and regular files and directories. Depending on your system settings files and directories, as well as hidden files and hidden directories, might appear in different colours.

Thus you can directly see what are files and what are directories. You may have noticed the directories "`.`" and "`..`". These are special directories. The directory "`.`" stands for the working directory. This is useful in commands like `cp` (copy). The "`..`" directory is the relative pathname to the parent directory. This relative pathname is helpful when you change to the parent directory with the command `cd` (change directory). The function of "`.`" and "`..`" is shown in Terminal 11 on page 40. Now we want to get some more information about the files and directories. We use the option `-l` (list). With this option the command `ls` lists the files and folders and gives additional information. First, in line 11, the size of the directory is shown. Here 12 Kilobyte are occupied. In line 15 we add the option `-h` (human). We obtain the same output, but human-readable. Now we see that the number 12 means 12K, which reads 12 Kilobytes. In line 20, we list the content of the directory *Text*. It contains the file *the.date*, which we saved at that location in line 5. With the option `-R` (recursively) you can list the content of subdirectories immediately (see Terminal 13 on page 45).

4.2 File Attributes

When you apply the command `ls -l`, you can see that any file or directory is connected to certain attributes. Let us take a closer look at line 13 of Terminal 10 on the page before in Fig. 4.2.

```
drwxrwxr-x    2     rw    rw    4.096   Apr 1 16:51   Text
 1      2         3     4     5      6            7          8
permissions nodes user group size      last change      name
file type
```

Fig. 4.2. Data obtained by the command `ls -l`. The *numbers in italics* correspond to the description in the text

The following list describes the individual file attributes in more detail. The numbers correspond to the italic numbers in Fig. 4.2.

1. **Type**
   The first character indicates whether the item is a directory (d), a normal file (-), a block-oriented device (b), a character-oriented device (c) or a link (l). Actually, a directory is a special type of file. The same holds for the special directories "`.`" and "`..`" which can be seen in line 9 in Terminal 10 on the preceding page.
2. **Access Modes**
   The access modes or permissions specify three types of users who are either allowed to read (r), write (w) or execute (x) the file. The first block

of 3 characters indicates your own rights, the next block of 3 characters the group's rights and the last block of 3 characters the right for any user. The dash character (-) means that the respective permission is not set. In line 17 in Terminal 10 on page 37 we see a directory (d). Everybody is allowed to read (r) and access (x) the directory. However, only the user *rw* and the group *rw* is allowed to create files (w) in that directory. In line 19 we see an ordinary file (-) which is not executable (x). In Section 4.4 on page 41 we will learn how the permissions can be changed.

3. **Number of Links**
   This item indicates the number of links or references to this file or folder. By default, there are two links to directory and one link to files. Any subdirectory increases the number of directory links by one.
4. **Owner**
   The user who created the file or directory. Later, the ownership can be changed with the command `chown`. Then the current owner is shown.
5. **Group**
   This item shows to which group the file or directory belongs. The group ownership can be changed with the command `chgrp`. In some Linux versions this item is not shown by default. In that case you have to use the option `-g` with the `ls` command.
6. **Size in Byte**
   The size of the file or directory. The exact number of bytes is shown. If the option `-h` (human) is used then the file size is rounded and the character K (Kilobyte) or M (Megabyte) is used. The size of a directory is that of the directory file itself, not of all the files in that directory. Furthermore, the size of the directory file is always a multiple of 1024, since directory entries are arranged in blocks of 1024 Byte (= 1 Kilobyte).
7. **Modification Date and Time**
   Here, the date and system time when the file was last modified is shown. For a directory, the date and time when the content last changed (when a file in the directory was added, removed or renamed) is shown. If a file or directory was modified more than 6 months ago, then the year instead of the time is shown.
8. **Name**
   Finally, the name of the file or directory is shown.

We will come back to certain aspects of file attributes later.

4.3 Special Files: . and ..

Each directory contains two special files named *dot* and *dotdot*. These files point to the directory and to the parent directory, respectively. Figure 4.3 on the next page illustrates this.

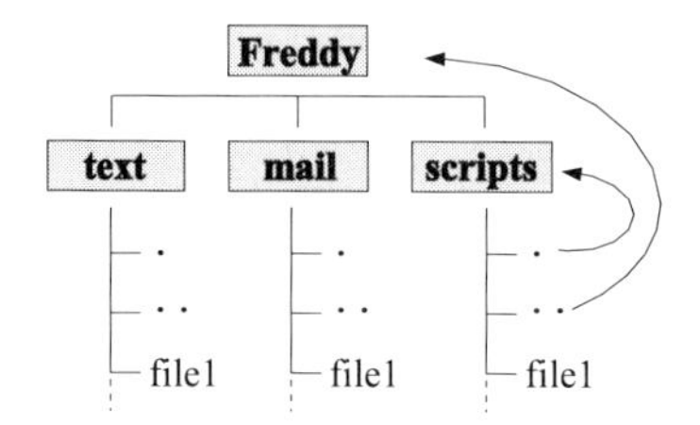

Fig. 4.3. The two special files *dot* and *dotdot* can be found in every directory. They are frequently used together with commands like `cd` (change directory) and `cp` (copy)

In Terminal 11 we learn more about the meaning of the directories "`.`" (dot) and "`..`" (dot dot).

```
Terminal 11: . and ..
1  $ pwd
2  /home/Freddy
3  $ mkdir temp
4  $ date > the.date
5  $ cd temp
6  $ pwd
7  /home/Freddy/temp
8  $ ls
9  $ cp ../the.date .
10 $ ls
11 the.date
12 $ cd ..
13 $ pwd
14 /home/Freddy
15 $ rm -r temp
```

Let us first check where we are. We use the command `pwd` (print working directory). Then we create the directory *temp* and the file *the.date* in the current directory */home/Freddy*. Next, in line 5, we change into the directory *temp* with the command `cd` (change directory). By using `pwd` we can prove that we changed into the directory *temp* which we created in line 3. With the `ls` command in line 8 we show that the directory is empty. Now we copy the file *the.date* which we created in directory */home/Freddy* into the currently active directory */home/Freddy/temp* (where we are right now). Therefore, we apply the command `cp` (copy). The parent directory to the currently active directory has the shortcut "`..`", whereas the current active directory has the shortcut "`.`". The syntax of the copy command is "`cp source destination`". Thus, with *../the.date* we call the file *the.date* which is one directory up. With the single dot we specify the directory we are in. This is the destination. Thus the whole command in line 9 copies the file *the.date* from */home/Freddy* to

/home/Freddy/temp. Note that you have duplicated the file. In order to move the file you have to substitute the command `cp` for the command `mv` (move). In line 12 we jump back to our home directory. Again we make use of the "`..`" shortcut. Then, in line 15, we remove the directory *temp* and all its content. Therefore we apply the command `rm` (remove) with the option `-r` (recursively). We can use only `rm` if we want to erase one or more files. However, to erase directories, even empty ones, you will need the option `-r`. Why is that? Well, we just saw that even an empty directory contains two, though hidden, files: "`.`" and "`..`".

4.4 Protecting Files and Directories

We learned already in Terminal 10 on page 37 that one characteristic of Linux files is ownership. Not everybody can do everything with files. This is an immense advantage since it makes the system safe. Note: You must be aware that the system administrator can access all files and directories. Thus, you should never keep secret information on a foreign computer.

4.4.1 Directories

Access permissions to directories help to control access to the files and subdirectories in that directory. If a directory has *read permission* (`r`), a user can run the `ls` command in order to see the directory content. He can also use wildcards (see Sect. 7.10 on page 91) to match files in this directory. If a directory has *write permission* (`w`), users can add, rename and delete files in that directory and all subdirectories. To access a directory, that is to read, write or execute files or programs, a user needs *execute permission* (`x`) on that directory. Note: To access a directory, a user must also have execute permission to all of its parent directories, all the way up to the root directory "/".

4.4.2 Files

Access permissions on a file always refer to the file's content. As we saw in the last paragraph, the access permissions of the directory where the file is located control whether the file can be renamed or removed. *Read permissions* (`r`) and *write permissions* (`w`) control whether you can read or change the file's content, respectively. If the file is actually a program it must have *execute permission* (`x`) in order to allow a user to run the program.

4.4.3 Examples

Let us take a look at some examples in order to clarify the situation. Look back at Terminal 10 on page 37 line 14. The first package of information, consisting of 10 characters, gives us some information about the file. First, indicated by the leading dash `-rw-r--r--`, we learn that the file is indeed a file and not a directory or something else. Now, let us look at the first block of permissions: `-rw-r--r--`. This block indicates the file owner's rights. The next block `-rw-r--r--` indicates the group's right and the last block `-rw-r--r--` indicates the access rights of all users. The following list exemplifies a number of different file attribute combinations.

`- rwx rwx rwx`	A file which can be read and modified by everybody. This setting gives no security at all.
`- rw- r-- r--`	This represents the standard setting for a file. The file can be read and modified by the owner, but only be read by all the others.
`- r-- --- ---`	This is the most secure setting. The owner can read but not modify the file. All the others can neither read nor modify the file.
`- rwx r-x r-x`	The standard setting for executable files. Everybody is allowed to run the program but only the user can change the file.
`d rwx r-x r-x`	The standard setting for directories. The user is allowed to access the directory and write and delete entries. All the other users can access the directory and list its content. However, they are not allowed to write or delete entries.
`d rwx --x --x`	Here the user has the same rights as in the example above. Other users are allowed to access the directory (with the command `cd`); however, they cannot list the content with the command `ls`. If the name of entries is known to the other users, they can, depending on the access rights of these entries, be read, written and executed.

4.4.4 Changing File Attributes

Now, in Terminal 12, we will play with file associations. Less common are the commands "`chown` *`user file`*" and "`chgrp` *`group file`*". They are used to change the user and the group ownership of a file or directory, respectively. The former commands can be executed only by the superuser root. We will concentrate on the command "`chmod` *`mode file`*".

Terminal 12: Ownership

```
1  $ id
2  uid=502(Freddy) gid=502(Freddy) Groups=502(Freddy)
3  $ date > thedate
4  $ ls -l
```

```
5  -rw-rw-r--  1  Freddy  Freddy  30 Apr 15 21:04 thedate
6  $ chmod g-w,o-r,a+x thedate
7  $ ls -l
8  -rwxr-x--x  1  Freddy  Freddy  30 Apr 15 21:04  thedate
9  $ chmod g=w thedate
10 $ ls -l
11 -rwx-w---x  1  Freddy  Freddy  30 Apr 15 21:04  thedate
12 $ chmod 664 thedate
13 $ ls -l
14 -rw-rw-r--  1  Freddy  Freddy  30 Apr 15 21:04  thedate
15 $
```

In line 1 of Terminal 12 we check our identity with the command `id`. We learn that Freddy has the user identity (uid) and group identity (gid) 502. The user Freddy belongs only to the group Freddy. In line 3 we generate the file *thedate*, which contains the current date. We do this only in order to have a file to play around with. Now let's check the file attributes with the command `ls -l` (list as list). Line 5 gives the result. For the meaning of the attributes look back to Section 4.2 on page 38. In line 6 we actively change the file attributes with `chmod`. In fact, we deny the group (`g`) access to modify (`w`) and others (`o`) to read (`r`) the file. All users (`a`) get the permission to execute (`x`) the file. Line 8 shows the changes. To change the access modes we state the user category, the type of change and the permissions to be changed. The letter code used for this command is shown in Table 4.2. Entries can be combined to something like *ug+rw*. Entries are separated by commas.

Table 4.2. Parameter that can be used with the command `chmod`

User	Type	Rights
u – user	+ add	r – read
g – group	- delete	w – write
o – others		x – execute
a – all		

Now, we apply another method to change file attributes. This is shown in line 9 in Terminal 12. Here the required access mode setting is directly entered. Again, settings can be combined and several entries are comma-separated, as in *ug=rw,a=x*. If you exclude the type, this is interpreted as nothing. With "`chmod go= filename`" the permissions are set to exactly nothing for group and others.

Finally, the last method we use is based on a number code. The generation of the code is shown in Table 4.3 on the following page. The numbers are allowed to lie between 0 and 7. For each user group (user, group, others) the code is calculated by adding the values for read (r=4), write (w=2) and execute (x=1) permission. If all permissions are required, the values add to 7, if only read and write permissions are to be set, the numbers add to 6 and so forth.

Table 4.3. How the number code is generated

User	Type	Rights
`r w x`	`r w x`	`r w x`
`4+2+1`	`4+2+1`	`4+2+1`
`7`	`7`	`7`

In line 12 of Terminal 12 on the page before we apply the code. With the command `chmod 664 thedate` we change the access permissions back to the default values: the user and users of the same group are allowed to read (`r`) and modify (`w`) the file *thedate*, whereas all the other users are only allowed to read the file.
If you want to change the permissions for many files at once you just list them one after the other: `chmod 777 file1 file2 file3`. If you use the wildcard "*" instead of file names, you change permissions for all files in the directory at once. If you use the number 0, you reset all permissions. For example, "`chmod 660 *`" sets the permissions of all files in the current directory to reading and writing for you and your group only.

4.4.5 Extended File Attributes

There is a new file attribute development on the way to getting established. It is developed by *Remy Card* and needs the *Linux second extended file system*. The corresponding program `chattr` (change attributes) largely expands the possibilities to restrict file and directory access. The letters "`ASacdijsu`" select the new attributes for the files: do not update time stamp (`A`), synchronous updates (`S`), append only (`a`), compressed (`c`), no dump (`d`), immutable (`i`), data journaling (`j`), secure deletion (`s`) and undeletable (`u`). For example, by making files append-only, their content can only be extended but not be deleted. Using the compressed (`c`) flag automatically compresses the file to save disk space. You have to check if you can use these extended file attributes on your system.

4.5 File Archives

File archives are frequently used when files are distributed via the internet or backed up onto a CD-ROM. In this section you will learn some basics on how to work with archived files.
From Windows you may be used to archive files with *WinZip*. This program creates a compressed file or even compressed directories. This is not possible

with Linux – at least not at the command line level. With Linux you first create an archive of files or directories and then compress the archive file. The most important command to create or extract an archive is `tar` (tape archive). The name already implies what the main purpose of the command is: to create a file archive for storage on a tape. Even nowadays, large data sets are stored on magnetic tapes. The reason is the high density of information that can be stored on magnetic tapes (maybe you also remember the *Datasette* from the *Commodore 64* times). Well, let us come back to the `tar` command. With `tar` you can put many directories and files into one single file. Thus, it is very easy to recover the original file organization. For that reason programs and other data are often distributed in the archived form. In your local working directory you then extract the archive and get back the original file and directory structure. As with many programs, in addition to the original `tar` program a GNU version called `gtar` is available on many systems. With `gtar` it is possible to compress files before they are put into the archive. By this, around 50% of memory can be saved (and, if you think further, money!). On most systems `gtar` completely substitutes for `tar`, although the command name `tar` has been kept. Thus, you do not have to bother about `gtar`. The output of `tar` can be sent either to a file or to a device like the floppy disk, a CD-ROM burner or a streamer.
Now let us use `tar`. First go to your home directory (`cd`) and create a directory called *tartest* (`mkdir`). Change into the newly created directory (`cd`) and create the file *tartest_file* with `date > tartest_file`. Then create the directories *tata1* and *tata2*. Within each of these two directories create 3 files named *tatax_filey* with the command `man tar > tata1_file1` (x is the number of the directory and y the file number). By redirecting (>) the output of the `man` command we create files with some content. You should end up with a list of files shown in Terminal 13.

Terminal 13: tar

```
1  $ ls -R
2  .:
3  tartest_file  tata1  tata2
4
5  ./tata1:
6  tata1_file1  tata1_file2  tata1_file3
7
8  ./tata2:
9  tata2_file1  tata2_file2  tata2_file3
10 $
```

In Terminal 13 we use the command `ls` with the option `-R` (recursively; note that the R is uppercase). With this command we list the content of the current working directory and the content of all its subdirectories. The output says that in the current working directory (”.” in line 2) we find the file *tartest_file* and the directories *tata1* and *tata2* (line 3). In the subdirectory

./tata1 (line 5) we find the files *tata1_file1* to *tata_file3* (line 6) and so on. Before we start using the `tar` command we should look at some of its options:

`c`	**C**reates a new archive.
`x`	**Ex**tracts files from an existing archive.
`t`	Prints a **t**able of contents of an existing archive. Do not forget to use the option f together with t! If you even add the option v, you will get a detailed content list with file sizes and so on. However, listing the content works only for uncompressed archives.
`f`	Expects a **f**ile *name* for the archive; the archive file will be named *name*. It is very helpful to end the file with *.tar*
`v`	**V**erbose, this means the program will print out what it is doing.
`z`	This option is available only with GNU-tar. It will compress the files after they have been added to the archive. Note that you have to decompress the archive with the same option in order to extract the files. It is very helpful to name compressed archives with the ending *.tgz*.

Of course, there are many more options. Take a look at the manpages (see Sect. 3.2.4 on page 33). As a big exception, the `tar` command takes the options without any dashes (`-`). Note that you should give your archives an appropriate filename extension: *.tar* for uncompressed and *.tgz* for compressed archives. Otherwise, you will not recognize the file as an archive! Now let us take a look at Terminal 14.

Terminal 14: tar

```
1  $ pwd
2  /home/Freddy/tartest
3  $ tar cvf ../daten.tar .
4  ./
5  ./tata1/
6  ./tata1/tata1_file1
7  ./tata1/tata1_file2
8  ./tata1/tata1_file3
9  ./tata2/
10 ./tata2/tata2_file1
11 ./tata2/tata2_file2
12 ./tata2/tata2_file3
13 ./tartest_file
14 $ tar cfz ../daten.tgz .
15 $ ls -lh ../daten*
16 -rw-rw-r-- 1 Freddy Freddy 100K Apr 19 15:55 ../daten.tar
17 -rw-rw-r-- 1 Freddy Freddy 5.7K Apr 19 15:55 ../daten.tgz
18 $
```

In line 1 we check that we are in the directory *tartest*. We then create (`c`) an archive with the filename (`f`) *daten.tar*, which will be written into the parent directory (`..`). The archive should contain everything which is in the current working directory (`.`). We choose to follow the progress of the program (`v`). In line 14 we create in the parent directory the archive *daten.tgz*, which is compressed (`z`). Furthermore, we do not wish to follow the program's progress (no verbose). In line 15 we use the command `ls` to list all files in the parent directory that begin with *daten*. Furthermore, we instruct `ls` to give file details (`-l`) and print the file size human-readable (`-h`). As you can see, the compressed archive is much, much smaller than the uncompressed version. Now take a look at the archives *daten.tar* and *daten.tgz* by typing `cat ../daten.tar |less` and `cat ../daten.tgz |zless`, respectively. You can scroll through the file content using ↑, ↓, PgUp and PgDn. In order to get back to the command line press Q.

How do we unpack an archive? Well, this is easy. Go to your home directory, where the files *daten.tgz* and *daten.tar* are, and make a directory called *tarout*. Then you have two possibilities to unpack the archive into *tarout*: either you change to the directory *tarout* and use "`tar xfz ../daten.tgz`" or you stay in your working directory and use "`tar xfz daten.tgz -C tarout/`". The `tar` command always extracts the archive into the present working directory, unless you explicitly name an existing output directory after the option `-C` (note that now the dash is required!). If you wish to extract the uncompressed archive *data.tar* you just omit the option `z`. It is very helpful to check the content of an archive with either "`tar ft archive_name`" for a short listing, or "`tar ftv archive_name`" for a detailed listing. This works only for uncompressed archives, for compressed archives you must add the option `z`!

4.6 File Compression

In order to save time, files distributed via the internet are usually compressed. There are several tools for file compression available. Most data-compression algorithms were originally developed as part of proprietary programs designed to compress files being interchanged across a single platform (e.g. PkZIP for MS-DOS and TAR for Unix). Data-compression modules have since developed into general purpose tools that form a key part of a set of operating systems. Two compression mechanisms, ZIP and TAR, have become ubiquitous, with a wide range of programs knowing how to make use of files compressed using these techniques. In Section 4.5 on page 44 we learned how to archive files using the command `tar`. Tar files are often compressed using the `compress` or `gzip` commands, with the resulting files having *.tar.z* (or *.taz* and *.tz*) or *tar.gz* (or *.tgz*) extensions, respectively (Table 4.4 on the next page).

`zip/unzip`

The most commonly used data-compression format is the ZIP format. ZIP files use a number of techniques to reduce the size of a file, including file shrinking, file reduction and file implosion. The ZIP format was developed by Phil Katz for his DOS-based program PkZIP in 1986 and is now widely used on Windows-based programs such as WinZip. The file extension given to ZIP files is *.zip*. The content of a zipped file can be listed with the command "`unzip -l` *`filename.zip`*".

`gzip/gunzip`

The GNU compression tool *gzip* is found on most Linux systems. Files are compressed by "`gzip` *`filename`*" and automatically get the extension *.gz*. The original file is replaced by the compressed file. After uncompressing using "`gzip -d` *`filename`*" or "`gunzip` *`filename`*" the compressed file is replaced by the uncompressed file. In order to avoid replacement you can use the option `-c`. This redirects the output to standard output and can thus be redirected into the file. The command "`gzip -c` *`filename`* `>` *`filename2`*" creates the compressed file *filename2* and leaves the file *filename* untouched. With the option `-r` you can recursively compress the content of subdirectories.
`gzip` is used when you compress an archive with `tar z` (see Sect. 4.5 on page 44). Then you get the file extension *.tgz*.

`bzip2/bunzip2`

The command `bzip2` is quite new and works more efficiently than `gzip`. Files are usually 20 to 30% smaller compared to `gzip` compressed files.

`compress/uncompress`

The compression program `compress` *`filename`* is quite inefficient and not widespread any more. Compressed files carry the extension *.Z*. Files can be uncompressed using `compress -d` *`filename`* (decompress) or `uncompress` *`filename`*.

Table 4.4 gives an overview of file extensions and their association with compression commands.

Table 4.4. File-compression commands and file extensions

Command	Extension(s)
`zip`	.zip
`compress`	.tar.z – .taz – .tz
`gzip`	.tar.gz – .tgz
`bzip2`	.bz2

4.7 Searching for Files

After a while of working, your home directory will fill up with files, directories and subdirectories. The moment will come that you cannot remember where you saved a certain file. The `find` program will help you out. It offers many ways in which to search for your lost file. `find` searches in directories for files or directories (remember that directories are a sort of file) that fulfil certain attributes. Furthermore, you can define actions to be executed when the file(s) are found. More than one attribute can be combined with a logical *and* or a logical *or*. Per default, `find` combines all search attributes with the logical *and*. To use *or*, the attributes must be separated by the `-o` and be enclosed in brackets. Note: The brackets must be preceded by an escape character. The syntax of the search command is `find` ***`directory attributes`***. Some important options are:

`-type`	Search for files (`f`) or directories (`d`)
`-name`	The name should be put into double quotes (" ") and may contain wildcards (*,?,[ ],...)
`-size`	Search for files larger (+) than a number of 512-byte blocks or characters (add a `c`). You can also search for files that must be smaller (-) than the given size.
`-ctime`	Search for files that were last changed x days ago. You can also search for files that are older (+) or younger (-) than x days.
`-exec`	Execute a command on all found files. The syntax is: "`command␣{}␣\;`". Notice the spaces! The curled brackets represent the found files.

Again, there are many more options available. You are always welcome to learn more about the commands by looking into the *manpages*, in this case by applying `man find` (see Sect. 3.2.4 on page 33). Let us perform a little exercise in order to get a better feeling for the `find` command. Go to your home directory (`cd`) and create a directory called *find-test* with the subdirectory *sub*. Now create two files within *find-test* (`date>find-test/1` and `date>find-test/2`) and one file within *find-test/sub* (`date>find-test/sub/3.txt`).

Terminal 15: find

```
1  $ mkdir find-test
2  $ mkdir find-test/sub
3  $ date>find-test/1
4  $ date>find-test/2
5  $ date>find-test/sub/3.txt
6  $ find find-test -type f -size -6 -name "[0-9]"
7  find-test/1
8  find-test/2
9  $ find find-test -type f -size -6 -name "[0-9]*"
```

```
10  find-test/1
11  find-test/2
12  find-test/sub/3.txt
13  $ find find-test -type f -size -6c -name "[0-9]*"
14  $ find find-test -type f -size -6 -name "[0-9]*"
15        -name "*.txt"
16  find-test/sub/3.txt
17  $ mkdir find-found
18  $ find find-test -type f -size -6 -name "[0-9]*"
19        -name "*.txt" -exec cp {} find-found \;
20  $ ls find-found/
21  3.txt
22  $
```

In Terminal 15 lines 1 to 5 we first create the directories and files as described above. In line 6 we search in the directory *find-test* for files (`-type f`) with a size less than 6 times 512 byte (`-size -6`) and a filename that contains any number from 0 to 9 (`-name [0-9]`). The brackets `[ ]` are wildcards (see Sect. 7.10 on page 91). The output is displayed in lines 7 to 8. Two files are found. In line 9 we allow the filename to be a number followed by any other character or none (`-name [0-9]*`). Again, the brackets `[ ]` and the star `*` are wildcards. Now, all three files are found. In line 13, we use the same attributes as in line 9, except that the file must contain less than 6 characters (`-size -6c`). There is no hit. Next, we search with the same options as above but the filename must contain any number from 0 to 9 followed by any other character or none (`*`) and end with *.txt* (`-name *.txt`). There is only one hit (*find-test/sub/3.txt*). In line 18/19 we use the same search pattern as in line 14/15, but apply the `cp` command to copy this file into the directory *find-found* (`-exec cp {} find-found \;`), which we created in line 17. A check with the `ls` command in line 20 shows us that we were successful.

Exercises

The following exercises should strengthen your file-handling power.

4.1. Go to your home directory and check that you are in the right directory. List all files in your home directory. List all the files in that directory including the hidden files. List all the files including the hidden files in the long format.

4.2. Determine where you presently are in the directory tree. Move up one level. Move to the root directory. Go back to your home directory.

4.3. Create a directory called *testdir* in your home directory, and a directory named *subdir* in the *testdir* directory (*testdir/subdir*). List all the files in the *testdir/subdir* subdirectory, including the hidden files. What are the entries you see? Remove the *subdir* subdirectory in your home directory.

4.4. Using the `cd` and the `ls` commands, go through all directories on your computer and draw a directory tree. Note: You may not have permission to read all directories. Which are these?

4.5. What permissions do you have on the */etc* directory and the */etc/passwd* file? List all the files and directories in your */home* directory. Who has which rights on these files, and what does that mean? Print the content of the file */etc/passwd* onto the screen.

4.6. In your home directory create a directory named *testdir2*. In that directory create a file containing the current date. Now copy the whole directory with all its content into the folder *testdir*. Remove the old *testdir2*. Check that you have successfully completed your mission.

4.7. Create the file *now* with the current date in the directory *testdir*. Change permissions of this file: give yourself read and write permission, people in your group should have read, but no write permission, and people not in your group should not have any permissions on the file.

4.8. What permissions do the codes 750, 600 and 640 stay for? What does `chmod u=rw,go-rwx myfile` do? Change permissions of all files in the *testdir* directory and all its subdirectories. You should have read and write permission, people in your group have only reading permission, and other people should have no permissions at all.

4.9. Create a new directory and move into it with a one-liner.

4.10. Use different compression tools to compress the manual page of the `sort` command. Calculate the compression efficiencies in percent.

4.11. Create the directory *testpress* in your home directory. Within *testpress* create 3 filled files (redirect manpages into the files). From your home directory use the commands `(un)compress`, `(un)zip`, `g(un)zip` and `b(un)zip2` to compress and uncompress the whole directory content (use the option `-r` (recursively)), respectively. What does the compressed directory look like?

4.12. Create a hidden file. What do you have to do in order to see it when you list the directories content?

4.13. Create a number of directories with subdirectories and files and create file archives. Extract the file archives again. Check whether files are overwritten or not.

4.14. Play around with copying, moving and renaming files. What is the difference between moving and renaming files? What happens to the file's time stamp when you copy, move or rename a file?

4.15. Create a directory in a directory, a so-called subdirectory. What happens to the number of links?

4.16. Create a file in a directory. Change this file. What happens to the modification date of the directory? Add a new file. What happens now? Rename a file with the command `mv` (move). What happens now?

4.17. Play around with the `find` command. Create directories, subdirectories and files and try out different attributes to find files.

5

Installing BLAST and ClustalW

In this chapter you will learn how to install small programs. As examples, we are using BLAST [1] (Basic Local Alignment Search Tool) and ClustalW [14]. BLAST is a powerful tool to find sequences in a database. Assume you have sequenced a gene and now want to check whether there are already similar genes sequenced by somebody else. Then you "blast" your sequence against an online database and get similar sequences, if present, as an output. Now assume you have found 10 similar genes. Of course, you would like to find regions of high similarity, that is, regions where these genes are conserved. For this, one uses ClustalW. ClustalW is a general-purpose multiple-sequence alignment program for DNA or protein sequences. It produces biologically meaningful multiple-sequence alignments of divergent sequences. ClustalW calculates the best match for the selected sequences and lines them up such that the identities, similarities and differences can be seen. Then evolutionary relationships can be visualized by generating cladograms or phylograms.
Both programs use different installation procedures. BLAST comes as a packed and compressed archive that needs only to be unpacked. ClustalW comes as a packed and compressed archive, too. However, before you can run the program it needs to be compiled. This means you get the source code of the program and must create the executable files from it.

5.1 Downloading the Programs via FTP

First of all you need to download the programs and save them in your home directory. How does this work? Well, of course, it is necessary that your computer is connected to the Internet! I assume that it is working for you.
Whenever you download a program you basically transfer data via the internet. This means you connect to a remote computer, copy a file to your computer and finally disconnect from the remote computer. There is a special program for this: *FTP*. FTP is the user interface to the *Internet standard **f**ile **t**ransfer **p**rotocol*, which was especially designed for file transfer via the

Internet. The program allows a user to transfer files to and from a remote network site. In fact, it is a very powerful program with many options. We will use only a small fraction of its capabilities. If you wish, take a look at the manual pages (`man ftp`).

5.1.1 Downloading BLAST

We will not download the newest version of BLAST, which is at the time of writing these lines version 2.2.8, but version 2.2.4. This version has less dependencies than the newer ones and should run on all typical Unix/Linux installations. Now, let us start. Go into your home directory by typing `cd` (remember: using the command `cd` without any directory name will automatically bring you to your home directory). Make a directory for BLAST by typing `mkdir blast`. Type `cd blast` to change into the *blast* directory. Then type "`ftp ftp.ncbi.nih.gov`". With this, you tell Linux to connect to the file server named *ftp.ncbi.nih.gov* using the file transfer protocol (*ftp*). Enter *anonymous* as the username. Enter your *e-mail address* as the password. Type `bin` to set the binary transfer mode. Next, type `cd blast/executables/release/2.2.4/` to get to the right directory on the remote file server and then type `ls` to see a list of the available files. You should recognize a file named *blast-2.2.8-ia32-linux.tar.gz*. Since we are working on a Linux system, this is the file we will download. If, instead, you work on a Unix workstation or with Mac OS X Darwin, you should download *blast-2.2.4-ia32-solaris.tar.gz* or *blast-2.2.4-powerpc-macosx.tar.gz*, respectively. To download the required file type the command `get` followed by the name of the file for your system, e.g. `get blast-2.2.8-ia32-linux.tar.gz`. Finally, type `quit` to close the connection and stop the ftp program. The file *blast-2.2.8-ia32-linux.tar.gz* should now be in your working directory. Check it using the command `ls`.

Terminal 16: Downloading BLAST

```
1  $ ftp ftp.ncbi.nih.gov
2  Connected to ftp.ncbi.nih.gov.
3      Public data may be downloaded by logging in as
4      "anonymous" using your E-mail  address as a password.
5  220 FTP Server ready.
6  Name (ftp.ncbi.nih.gov:rw): anonymous
7  331 Anonymous login ok,
8      send your complete email address as your password.
9  Password:
10 230 Anonymous access granted, restrictions apply.
11 Remote system type is UNIX.
12 Using binary mode to transfer files.
13 ftp> bin
14 200 Type set to I
15 ftp> cd blast/executables/release/2.2.4/
16 250 CWD command successful.
```

```
17 ftp> ls
18 500 EPSV not understood
19 227 Entering Passive Mode (130,14,29,30,197,39).
20 150 Opening ASCII mode data connection for file list
21 -r--r--r--   1 ftp      anonymous 15646697 Aug 29  2002
22     blast-2.2.4-ia32-linux.tar.gz
23 ...
24 -r--r--r--   1 ftp      anonymous 41198447 Aug 29  2002
25     blast-2.2.4-powerpc-macosx.tar.gz
26 226 Transfer complete.
27 ftp> get blast-2.2.4-ia32-linux.tar.gz
28 local: blast-2.2.4-ia32-linux.tar.gz
29     remote: blast-2.2.4-ia32-linux.tar.gz
30 227 Entering Passive Mode (130,14,29,30,196,216).
31 150 Opening BINARY mode data connection for
32     blast-2.2.4-ia32-linux.tar.gz (15646697 bytes)
33 226 Transfer complete.
34 15646697 bytes received in 02:53 (87.96 KB/s)
35 ftp> quit
36 221 Goodbye.
37 $
```

Terminal 16 gives an in-detail description of the `ftp` commands necessary in order to download BLAST. Furthermore you see the feedback from the remote computer [please note that the output had to be truncated (indicated by `"..."`) here and there for printing reasons]. In line 1 we apply the command to connect to the remote FTP server called *ftp.ncbi.nih.gov*. In line 6 we enter the login name *anonymous* and in line 9 the password, which is your email address. With the command `bin` in line 13 we change to the binary transmission mode (which is not absolutely necessary since FTP usually recognizes the file type automatically). Then we change to the directory that contains the executables (line 15) and list its content (line 17). Among others you will recognize a file called *blast-2.2.8-ia32-linux.tar.gz*, which we download with the command listed in line 27. Finally we quit the session in line 35. The installation process is described in Section 5.2 on page 57.

5.1.2 Downloading ClustalW

In contrast to BLAST, ClustalW does not come as a binary file, which can be directly executed after decompressing. Instead, you get the source code and have to compile the source code for your Unix/Linux system; but first let us download the source code at *ftp://ftp.ebi.ac.uk/pub/software/unix/clustalw*. The file we need to download is *clustalw1.83.UNIX.tar.gz*. Download the file into a directory called *clustal* (`mkdir clustal` and `cd clustal`).

Terminal 17: Downloading ClustalW

```
1 $ ftp ftp.ebi.ac.uk
2 Connected to alpha4.ebi.ac.uk.
```

```
3  ...
4  Name (ftp.ebi.ac.uk:rw): anonymous
5  331 Guest login ok,
6      send your complete e-mail address as password.
7  Password:
8  230-Welcome anonymous@134.95.189.5
9  ...
10 230 Guest login ok, access restrictions apply.
11 Remote system type is UNIX.
12 Using binary mode to transfer files.
13 ftp> cd pub/software/unix/clustalw
14 ...
15 ftp> get clustalw1.83.UNIX.tar.gz
16 local: clustalw1.83.UNIX.tar.gz
17 remote: clustalw1.83.UNIX.tar.gz
18 500 'EPSV': command not understood.
19 227 Entering Passive Mode (193,62,196,103,227,137)
20 150 Opening BINARY mode data connection for
21     clustalw1.83.UNIX.tar.gz (166863 bytes).
22 100% |****************|   162 KB  151.41 KB/s    00:00 ETA
23 226 Transfer complete.
24 166863 bytes received in 00:01 (148.60 KB/s)
25 ftp> pwd
26 257 "/pub/software/unix/clustalw" is current directory.
27 ftp> !pwd
28 /home/emboss/Freddy/clustal
29 ftp> !ls
30 clustalw1.83.UNIX.tar.gz
31 ftp> quit
32 221-You have transferred 166863 bytes in 1 files.
33 ...
34 221 Goodbye.
35 $
```

Terminal 17 shows details about the FTP session. In addition to what we did in Terminal 16 on the preceding page, we play a little bit with shell commands. In line 25 we use the command `pwd` to print the working directory of the remote computer. If we wish to execute a shell command on our local machine we need to precede it with the exclamation mark (`!`). Thus, in line 27 we print the working directory of our computer. We then check its content with `!ls` and can confirm that the file *clustalw1.83.UNIX.tar.gz* has been downloaded successfully. The installation process itself will be explained in Section 5.4 on page 59.

5.2 Installing BLAST

Be sure you are in the directory *blast* and that you correctly downloaded the file *blast-2.2.8-ia32-linux.tar.gz* into it as described in Section 5.1.1 on page 54. Now, type

```
gunzip blast-2.2.8-ia32-linux.tar.gz
```

Next, type

```
tar -xf blast-2.2.8-ia32-linux.tar
```

This will extract the archive into the current working directory (see Sect. 4.5 on page 44). You are done! BLAST is distributed with ready-to-go executable files. You can run the program directly. To run, for example, the program `blastall` you just type in the current directory `./blastall`. Why do you have to precede the command with "`./`"? Well, the program is not registered in the path of the system. The path is a variable that contains all directories where the system searches for executable programs (see Sect. 8.2 on page 99). By typing `./` you explicitly define the location of the command. From the parent directory of *blast* you would have to call the program with `blast/blastall`. Of course, `blastall` does not do too many interesting things at this stage since we have not provided any data for it. Instead, the program displays all its options. However, it demonstrates that the program is alive.

5.3 Running BLAST

Now, let us see how to run BLAST. First, we have to create a sequence database we query against. Usually, this would be a whole genome. However, for our test, we generate a hypothetical mini-genome. For this purpose we use the command `cat`. The normal function of this command is to show or concatenate file content. However, we can misuse the `cat` command to create a text file, as shown in the following Terminal (see also Sect. 6.1 on page 64).

Terminal 18: BLAST Search

```
1  $ cat >testdb
2  >seq1
3  atgcatgatacgatcgatcgatcgatgcatcgatcg
4  >seq2
5  tagctagcatcgcgtcgtgctacgtagctcgtgtgtatgtg
6  $ cat >testblast
7  >search
8  cgatcgatcgat
9  $ ./formatdb -i testdb -p F
10 $ ls testdb*
11 testdb  testdb.nhr  testdb.nin  testdb.nsq
12 $ ./blastall -p blastn -d testdb -i testblast
```

```
13  BLASTN 2.2.4 [Aug-26-2002]
14  Query= search (12 letters)
15  Database: testdata
16             2 sequences; 77 total letters
17  >seq1
18            Length = 36
19   Score = 24.3 bits (12), Expect = 3e-05
20   Identities = 12/12 (100%)
21   Strand = Plus / Plus
22  Query: 1  cgatcgatcgat 12
23            ||||||||||||
24  Sbjct: 15 cgatcgatcgat 26
25
26   Score = 24.3 bits (12), Expect = 3e-05
27   Identities = 12/12 (100%)
28   Strand = Plus / Plus
29  Query: 1  cgatcgatcgat 12
30            ||||||||||||
31  Sbjct: 11 cgatcgatcgat 22
32
33   Score = 24.3 bits (12), Expect = 3e-05
34   Identities = 12/12 (100%)
35   Strand = Plus / Minus
36  Query: 1  cgatcgatcgat 12
37            ||||||||||||
38  Sbjct: 24 cgatcgatcgat 13
39  $
```

In line 1 of Terminal 18 we create an empty file called *testdb*. After you have entered the command `cat >testdb` and hit (Enter) you end up in an empty line. Everything you enter now will be redirected into the file *testdb*. However, you can correct errors with the (BkSp) key. When you hit (Enter), a line break will be generated. You finish by pressing (Ctrl)+(D) at the end of line 5. Next, you create the text file *testblast* which contains the query sequence you want to "blast" against the database (i.e. you search similar sequences in the database). In line 9 we generate files which are used by BLAST and which are specific for our database file *testdb*. This is done with the command

```
./formatdb -i testdb -p F
```

The first option (`-i testdb`) identifies the input database file, here *testdb*, which we created earlier. The second option (`-p F`) tells BLAST that the database contains nucleic acid and not protein sequences. Remember that you have to precede the command `formatdb` with `./` because it is not a system command and not yet in the system command path. Let us see what kind of files have been generated. All new files created by `formatdb` start with the database filename we provided (*testdb*). This means we can specifically search for new files using `ls testdb*` as shown in line 10. We should see 3 new files.

Now the interesting part starts! We query the database for our query sequence, which is in the file called *testblast*. Therefore, we apply the command

```
./blastall -p blastn -d testdb -i testblast
```

The first option (`-p blastn`) defines the query program. With `blastn` we choose the normal program to query nucleotide sequences. Furthermore, we specify the name of the database (*testdb*) and the name of the file containing the query sequence (*testblast*). (You could also enter several sequences into the query file. They would be processed one after the other. It is, however, important that you use the *FASTA format* for the file.) The output starts in line 13. Note that, for printing reasons, the output in Terminal 18 on the preceding page is shorter than in reality. The sequence was actually found three times in sequence 1.

5.4 Installing ClustalW

After we have successfully downloaded the compressed source file for ClustalW, we need to uncompress and compile it. From the file extension *.tar.gz* we see that the compression program `gzip` was used (see Sect. 4.6 on page 47) and that the file is an archive (see Sect. 4.5 on page 44). One way to unzip the files and extract the archive is

```
gunzip clustalw1.83.UNIX.tar.gz
```

followed by

```
tar -xf clustalw1.83.UNIX.tar
```

A new directory named *clustalw1.83* will be created. Change into this directory (`cd clustalw1.83`). Now we start the compilation process: just type `make`. You will see a number of lines of the type *cc -c -O interface.c* popping up. Finally you get back to your shell prompt. When you now list (`ls`) the directory's content, you will recognize a number of new files with the file extension *.o* and one new executable file called `clustalw`. That is the program compiled for your system! How do you recognize an executable file? Take a look back into the section on file attributes (see Sect. 4.2 on page 38).

5.5 Running ClustalW

After you have successfully compiled ClustalW we should check whether it runs or not. We will align three tRNA genes. These are pretty short genes; but still we save ourselves from typing and restrict the sequences to only 25 nucleotides.

Terminal 19: ClustalW

```
1  $ cat>tRNA
2  >gene1
3  GGGCTCATAGCTCAGCGGTAGAGTG
4  >gene2
5  GGGCCCTTAGCTCAGCTGGGAGAGA
6  >gene3
7  GGGGGCGTAGCTCAGCTGGGAGAGA
8  $ ./clustalw tRNA
9  ...
10 $ cat tRNA.aln
11 CLUSTAL W (1.83) multiple sequence alignment
12
13 gene2           GGGCCCTTAGCTCAGCTGGGAGAGA-
14 gene3           GGGGGCGTAGCTCAGCTGGGAGAGA-
15 gene1           GGGCTCATAGCTCAGC-GGTAGAGTG
16                 ***  * ********* ** ****
17 $
```

In Terminal 19 we first create a file called *tRNA*, which contains the sequences we want to align. Therefore, we apply the command `cat` as explained in Section 6.1 on page 64. In line 8 of Terminal 19 we actually start the program ClustalW with

`./clustalw tRNA`

We get a lot of output on the screen, which is omitted here. The alignment is not printed out onto the screen but written into the file *tRNA.aln*. We can take a look at the alignment using `cat tRNA.aln` (see line 10). There are many more options for ClustalW available. Moreover, in addition to the alignment, a phylogenetic tree is created; but this is not our focus here.

Exercises

In these exercises you will download and compile the program `tacg` [8]. In order to do so, you require to be connected to the internet. If you succeed, you can already consider yourself as a little Linux freak. `tacg` is a command-line program that performs many of the common routines in pattern matching in biological strings. It was originally designed for restriction enzyme analysis and while this still forms a core of the program, it has been expanded to fulfil more functions. `tacg` searches a DNA sequence read from the command line or a file for matches based on descriptions stored in a database of patterns. These descriptions can, e.g., be formatted as explicit sequences, matrix descriptions or regular expressions describing restriction enzyme cutting sites, transcription factor binding sites or what ever. The query result is sent to the standard output (the screen) or a file. With `tacg` you can also translate DNA sequences to protein sequences or search for open reading frames in any frame. It is approximately 5–50 times faster than the comparable routines in Emboss (see Sect. 2.10.1 on page 21). For more details you should take a look at (*tacg.sourceforge.net*).

5.1. Connect to the FTP server *ftp.sunet.se*, login anonymously and change to the following directory: *pub/molbio/restrict-enz/*. List the directory content and download the file *tacg-3.50-src.tar.gz* into your home directory. Log off and move the file into a directory called *dna_grep*.

5.2. Unzip and extract the archive contained in *tacg-3.50-src.tar.gz*.

5.3. In this exercise you perform the compilation step. Follow the instructions given in the file *dna_grep/tacg-3.50-src/INSTALL*. Check if you were successful by typing `./tacg`.

5.4. If you wish, you can take a look at the documentation in the subfolder *Docs* and play around a little.

6

Working with Text

After learning some basics in the previous chapter, you will learn more advanced tools from now on. Since you are going to learn programming, you should be able to enter a program, that is, a text. Certainly your programs will not run immediately. This means you need to edit text files, too. In this chapter you will use very basic text-editing tools. We will concentrate on the text editor `vi`.
With Linux you have the choice from an endless list of text editors. From Microsoft Windows you might have heard about the *Notepad*. That is the standard Windows text editor. With Linux you can choose between `ed`, `vi`, `vim`, `elvis`, `emacs`, `nedit`, `kedit`, `gedit` and many more. In general, we can distinguish between three different types of text editors: a) One group of text editors is line-orientated. `ed` belongs to this group. You can only work on one line of the text file and then need a command to get to the next line. This means that you cannot use the arrow keys to move the cursor (your current position) through the text. This is stone age text editing and good only for learning purposes. b) The next group of text editors is screen-oriented. `vi` belongs to this group of editors. You see the text over the whole screen and can scroll up and down and move to every text position using the arrow keys. This is much more comfortable than using a line-oriented editor. We will predominantly work with `vi`. It is powerful and comfortable enough for our purpose and usually available on all systems. You can even use it when you have no X-Server (see Sect. 2.3 on page 14) running, which is often the case when you login to a remote computer. c) The most comfortable text editors, of course, use the X-Server. However, you should not confuse comfortable with powerful! `nedit` or `kedit` are examples of editors which use the graphical user interface (GUI). Here you can use the mouse to jump around in the text and copy and paste (drag and drop) marked text. We are not going to work with these nice editors.

6.1 A Quick Start: cat

Now, let us start with a very easy method to input text. For this, we use the command `cat` (concatenate). We have already used `cat` for displaying the content of a text file (see for example Terminal 5 on page 30). We have also learned about redirecting the output of command into a file by using `>`. Now, when you use `cat` without specifying a filename you redirect the standard input (the keyboard) to the standard output (the screen). Let us try it out.

Terminal 20: cat

```
1 $ cat
2 this is text, now press Enter
3 this is text, now press Enter
4 now press Ctrl-Dnow press Ctrl-D now press Enter AND Ctrl-D
5  now press Enter AND Ctrl-D
6 $
```

In Terminal 20 we first run the command `cat` without anything. After hitting (Enter) we get into line 2 and enter the text "this is text, now press Enter". At the end of line 2 we press (Enter). The text we just entered will be printed again and we end up in line 4. Here we enter "now press Ctrl-D" and then press (Ctrl)+(D). Again, the text will be printed, but now we end up in the same line. Now we press first (Enter) and then (Ctrl)+(D). We are back. What happens is that the input from the keyboard is redirected to the screen. When we press (Enter) a new line starts before the redirection; but this is not important now. It is important that you realize that we could also redirect the standard input into a file! How?

Terminal 21: cat

```
1 $ cat > text-file
2 bla bla. Now press Enter
3 ahh, a new line! Now press Ctrl-D$ cat text-file
4 bla bla. Now press Enter
5 ahh, a new line! Now press Ctrl-D$
6 $ rm text-file
7 $
```

In Terminal 21 we redirect the standard input into a file named *text-file*. In line 2 we enter some text and then hit (Enter). We end up in line 3, enter some text again and press (Ctrl)+(D). We get back the command line directly. No new line is generated. This is uncomfortable! If we now, as shown in lines 3 to 5, display the text with `cat text-file`, we will see the same phenomenon in line 5: the dollar character is at the right of this line, instead of being at a new line. However, if you now press (Enter) you get into a new line. Hint: You should always finish a text file with a new line. Finally, in line 6 we remove the file that we have just created with the command `rm`.
It is easy to create a text file, isn't it? As shown in Terminal 22, you should

now create a text file named *genomes.txt*, which we are going to use and play around with. Remember: When you mistype something you can go back with the (BkSp) key – but only in the active line!

Terminal 22: Working File

```
1  $ cat>genomes.txt
2  H. sapiens (human) - 3,400,000,000 bp - 30,000 genes
3  A. thaliana (plant) - 100,000,000 bp - 25,000 genes
4  S. cerevisiae (yeast) - 12,100,000 bp - 6034 genes
5  E. coli (bacteria) - 4,670,000 bp - 3237 genes
6  4 lines text?
7  4 lines text?
8  203 characters?
9  $
```

Now let us see what we can do with the text file. We have already learned something about sorting the content of a text file in Terminal 7 on page 31. Let us take a look at it again.

6.1.1 Text Sorting

With `sort` you can, as the name implies, sort the content of a text file. This can be very helpful in order to make files more readable. The `sort` command comes with some useful options, which are, as usual, separated by a dash.

- `-n` Sorting takes into account the value of numbers. Otherwise, only their position in the alphabet would be considered and 50 would be printed before 8.
- `-r` The sorting result is displayed in reverse order.
- `-f` Lower- and uppercase characters are treated equally.
- `-u` Identical lines are printed only once. Thus you get unique or non-redundant output.
- `-k x` Sorts according to the content of field *x*. For example, "`-k 3`" sorts according to the content of field 3.
- `-t s` Set *s* as a new field separator (instead of the space character). For example, "`-t :`" sets the colon and "`-t \t`" sets the tabulator as field separator, respectively.

How does this work in real life?

Terminal 23: Text Sorting

```
1  $ sort -nu genomes.txt
2  H. sapiens (human) - 3,400,000,000 bp - 30,000 genes
3  4 lines text?
4  203 characters?
5  $ sort -n genomes.txt | sort -u
6  203 characters?
```

```
 7  4 lines text?
 8  A. thaliana (plant) - 100,000,000 bp - 25,000 genes
 9  E. coli (bacteria) - 4,670,000 bp - 3237 genes
10  H. sapiens (human) - 3,400,000,000 bp - 30,000 genes
11  S. cerevisiae (yeast) - 12,100,000 bp - 6034 genes
12  $ sort -u genomes.txt | sort -n
13  A. thaliana (plant) - 100,000,000 bp - 25,000 genes
14  E. coli (bacteria) - 4,670,000 bp - 3237 genes
15  H. sapiens (human) - 3,400,000,000 bp - 30,000 genes
16  S. cerevisiae (yeast) - 12,100,000 bp - 6034 genes
17  4 lines text?
18  203 characters?
19  $  sort -u genomes.txt | sort -n > sorted-genomes.txt
20  $
```

In Terminal 23 we use the options `-n` and `-u` in order to sort the content of the *genomes.txt* file, take care of correct sorting of numbers and get rid of doublets; but what is this? The output of `sort -nu genomes.txt` in lines 2 to 4 is a little bit short! Three organisms are missing. Okay, here we encounter an error (*bug*) in the program `sort`. With such a well-established command as `sort`, this is a rather rare situation. What can we do? First of all you should report the bug to the Linux community and thereby help to improve the commands. In the manpages (`man sort`) there is a statement where to report bugs: "Report bugs to bug-textutils@gnu.org". Then you have to think about alternatives to your original task. In our case the alternative is shown in line 5. Here we learn a new way of redirecting the output of a program: *piping*. Pipes are powerful tools to connect commands (see Sect. 7.6 on page 87). We have already seen how to redirect the output from a command into a file (`>`). Now we redirect the output from one command to another. Therefore, we use the "`|`" character. The "`|`" is a pipe, and this type of pipe sends the stream of data to another program. The use of pipes means that programs can be very modular, each one executing some narrow, but useful task. In Terminal 23 line 5 we pipe the output of the `sort` command with the option `-n` (sort numbers according to their value) to the `sort` command with the option `-u` (eliminate duplicates). The result will be displayed on the screen. We could also redirect the result into a file, as shown in line 19.
The next example uses the options "`-t`" and "`k`" in order to define a new field delimiter and the field, which is to be considered for sorting.

Terminal 24: Field Directed Sorting

```
1  $ sort -t - -k 3 -n genomes.txt
2  203 characters?
3  4 lines text?
4  4 lines text?
5  A. thaliana (plant) - 100,000,000 bp - 25,000 genes
6  H. sapiens (human) - 3,400,000,000 bp - 30,000 genes
7  E. coli (bacteria) - 4,670,000 bp - 3237 genes
```

```
8 S. cerevisiae (yeast) - 12,100,000 bp - 6034 genes
9 $
```

With the `sort` command in line 1 of Terminal 24 the file *genomes.txt* is sorted according to the contents of field 3 (`-k 3`). The field delimiter is set to the dash character (`-t -`). With these options, `sort` becomes very versatile and can be used to sort tables according to the values of a certain column.

6.1.2 View File Beginning or End

Imagine you have a number of text files but you cannot remember their content. You just want to get a quick idea. You could use `cat` or a text editor, but sometimes it is enough to see just a couple of lines to remember the file content. `head` and `tail` are the commands of choice. `head filename` displays the first 10 lines of the file *filename*. `tail filename` does the same for the last 10 lines. You might want to use the option `-n`. The command

```
head -n 15 structure.pdb
```

displays the first 15 lines of the file *structure.pdb*. Of course, the `-n` option applies to `tail`, too.

6.1.3 Scrolling Through Files

Assuming that you are an eager scientist collecting lots of data and having accordingly large files, you might look for something else than `cat`, `head` or `tail`. Of course, you want to see the whole file and scroll through it! If you want to read a long file at the screen, you should use `less`. (There is another program called `more`. It is older and thus not powerful. Just to let you know!). The syntax is easy:

```
less filename
```

With `less` you can scroll forward and backward through the text file named *filename*. After invoking `less`, the next page is displayed with (Space) and the next line is displayed with (Enter). By pressing (Q) you quit `less`. There are two nice commands that let you search forward and backward through an opened text file: `/word` and `?word`, respectively.
On some systems you will find the command `zless`, which can be used in order to view compressed files.

6.1.4 Character, Word and Line Counting

Another thing we might want to do is counting the number of lines, words and characters of a text file. This task can be performed with the command `wc` (word count). `wc` counts the number of bytes, whitespace-separated words and lines in each given file.

Terminal 25: wc

```
$ wc genomes.txt
       7      44     247 genomes.txt
$
```

As we can see from the output in Terminal 25, the file *genomes.txt* consists of 7 lines, 44 words and 247 characters.

6.1.5 Finding Text

A very important command is `grep` (global regular expression parser). `grep` searches the named input file(s) for lines containing a match to a given pattern. By default, `grep` prints the matching lines.

Terminal 26: grep

```
$ grep human genomes.txt
H. sapiens (human) - 3,400,000,000 bp - 30,000 genes
$ grep genes genomes.txt | wc -w
     36
$ grep genes genomes.txt | wc -l
      4
$ grep -c human genomes.txt
1
$ grep -c genes genomes.txt
4
$ grep 'H. sapiens' genomes.txt
H. sapiens (human) - 3,400,000,000 bp - 30,000 genes
$ grep 'bacteria\|human' genomes.txt
H. sapiens (human) - 3,400,000,000 bp - 30,000 genes
E. coli (bacteria) - 4,670,000 bp - 3237 genes
$
```

`grep` really is much more powerful than you can imagine right now! It comes with a whole bunch of options, only one of which was applied in Terminal 26. In line 1 we search for the occurrences of the word *human* in the file *genomes.txt*. The matching line is printed out. In line 3 we search for the word *genes* and pipe the matching lines to the program `wc` to count the number of words (option `-w`) of these lines. The result is 36. Next, in line 5, we count the number of lines (`wc` option `-l`) in which the word *genes* occurs; but we can obtain this result much more easily by using the `grep` option `-c` (count). With this option `grep` displays only the number of lines in which the query expression occurs. How can we search for two consecutive words? This is shown in line 12. Notice that we enclose the words we are searching for in single quotes (you could also use double quotes); and how about querying for two words that are not necessarily in one line? Use the combination "`\|`" between the words as shown in line 13 of Terminal 26. This stands for the logical *or*. Be careful not to include spaces!

At this stage you should already have a feeling about the power of the Linux command line. We will get back to `grep` when we talk about regular expressions in Chapter 9 on page 127.

6.1.6 Text File Comparisons

Two interesting commands to compare the contents of text files are `diff` and `comm`. To see how these commands work, create two text files with a list of words, as shown in the following Terminal.

Terminal 27: diff

```
1  $ cat>amino1
2  These amino acids are polar:
3  Serine
4  Tyrosine
5  Arginine
6  $ cat>amino2
7  These amino acids are polar and charged:
8  Lysine
9  Arginine
10 $ diff amino1 amino2
11 1,3c1,2
12 < These amino acids are polar:
13 < Serine
14 < Tyrosine
15 ---
16 > These amino acids are polar and charged:
17 > Lysine
18 $ diff -u amino1 amino2
19 --- amino1    2003-05-11 18:42:53.000000000 +0200
20 +++ amino2    2003-05-11 18:43:30.000000000 +0200
21 @@ -1,4 +1,3 @@
22 -These amino acids are polar:
23 -Serine
24 -Tyrosine
25 +These amino acids are polar and charged:
26 +Lysine
27  Arginine
28 $ diff -c amino1 amino2
29 *** amino1    2003-05-11 18:42:53.000000000 +0200
30 --- amino2    2003-05-11 18:43:30.000000000 +0200
31 ***************
32 *** 1,4 ****
33 ! These amino acids are polar:
34 ! Serine
35 ! Tyrosine
36   Arginine
37 --- 1,3 ----
```

```
38 ! These amino acids are polar and charged:
39 ! Lysine
40   Arginine
41 $
```

The result of the command `diff` (difference) indicates what you have to do with the file *amino1* to convert it to *amino2*. You have to delete the lines marked with "<" and add the lines marked with ">". With the option `-u` (line 18) the same context is shown in another way. The option `-c` (line 28) is used to display the differences only (indicated by a "`!`").
The command `comm` (compare) requires that the input files are sorted. We do this in line 1 in Terminal 28. Note: We can write several commands in one line, separating them with the semicolon character (`;`).

Terminal 28: comm

```
1  $ sort amino1>amino1s; sort amino2>amino2s
2  $ comm amino1s amino2s
3                  Arginine
4          Lysine
5  Serine
6  These amino acids are polar:
7          These amino acids are polar and charged:
8  Tyrosine
9  $ comm -12 amino1s amino2s
10 Arginine
11 $
```

The `comm` command prints out its result in three columns. Column one contains all lines that appear only in the first file (*amino1s*), column two shows all lines that are present only in the second file (*amino2s*) and the third column contains all files that are in both files. You can restrict the output with the options `-n`, n being one or more column numbers, which are not to be printed. Line 9 in Terminal 28 displays only the content of the third column ("minus 1 and minus 2"), which contains the common file content.

6.2 pico

Probably the easiest-to-handle text editor is called `pico` (**pi**ne **co**mposer). It has the same look and feel as an old DOS editor (see Fig. 6.1 on the next page). However, `pico` is not installed on all Unix/Linux systems. Thus, you should make the effort to learn `vi` (see Sect. 6.3 on page 72), which is a bit harder but more universal and which offers you many more possibilities (in fact, you need to memorize only 6 commands in order to work `vi`). The `pico` editor was developed by the University of Washington. You start `pico` with the command `pico` and, optional, with a filename. If the file already exists,

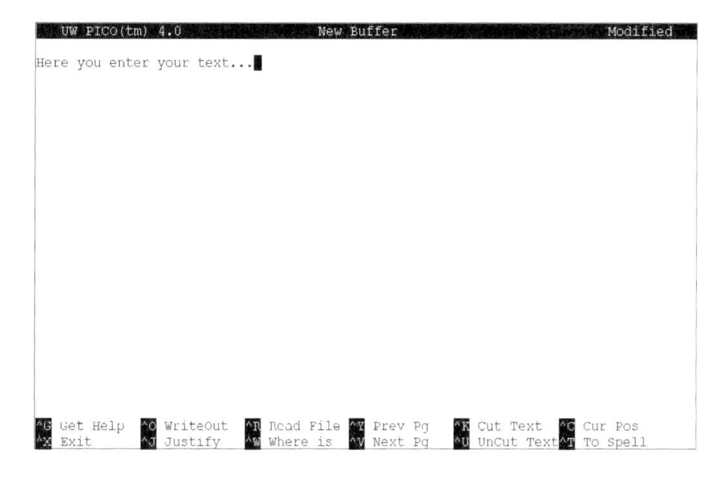

Fig. 6.1. The main window of the text editor `pico`

it will be opened. When you write a program it is highly recommended to use the option `-w` (wrap). It tells `pico` not to break (wrap) the lines at the right margin of the editing window, but only when you press the (Enter) key. Thus, the command to start `pico` would look like this: `pico -w filename`. If you provide no filename, that is fine as well. You can save your text later and provide a filename at that time.

When you have started `pico` you will get a kind of a graphical interface (see Fig. 6.1). The top of the display, called status line, shows the version number of `pico` that you use, the current file being edited and whether or not the file has been modified. The third line from the bottom is used to report informational messages and for additional command input. Now comes the comfortable part of `pico`: the bottom two lines list the available editing commands. Depending on your current action, the content of this window might change.

Each character you type is automatically inserted at the current cursor position. You can move the cursor with the arrow keys. If this does not work, you have to use the following commands:

(Ctrl)+(F)	move **f**orward one character
(Ctrl)+(B)	move **b**ackward one character
(Ctrl)+(P)	move to the **p**revious line
(Ctrl)+(N)	move to the **n**ext line
(Ctrl)+(A)	move to the beginning of the current line
(Ctrl)+(E)	move to the **e**nd of the current line
(Ctrl)+(V)	move forward a page of text
(Ctrl)+(Y)	move backward a page of text

Editing commands are given to `pico` by typing special control key sequences. A circumflex, ^, is used to denote the Ctrl key. Thus, "^X Exit" translates to: press Ctrl+X to exit the program. You can press Ctrl+G to get a help text. This help text gives you more information about all available commands.

6.3 `vi` and `vim`

The text editor `vi` (**vi**sual, or **vi**rtual editor) is an improvement on the original Unix editor `ed`. Nowadays it is usually replaced by `vim` (**vi im**proved) (see Fig. 6.2), which is executed with the command `vi`, too.

```
~
~
~
~
~
~
~
~                          VIM - Vi IMproved
~
~                          version 6.0z ALPHA
~                      by Bram Moolenaar et al.
~              Vim is open source and freely distributable
~
~                     Help poor children in Uganda!
~             type  :help iccf<Enter>       for information
~
~             type  :q<Enter>               to exit
~             type  :help<Enter>  or  <F1>  for on-line help
~             type  :help version6<Enter>   for version info
~
~
~
~
~
~
```

Fig. 6.2. The text editor `vim`, which nowadays replaces `vi`

`vi` (pronounced vee-eye) provides basic text-editing capabilities. Three aspects of `vi` make it appealing. First, `vi` is supplied with all Unix systems. Thus, you can use `vi` at other universities or any businesses with Unix or Linux systems. Second, `vi` uses a small amount of memory, which allows efficient operation when the network is busy. Third, because `vi` uses standard alphanumeric keys for commands, you can use it on virtually any terminal or workstation in existence without having to worry about unusual keyboard mappings. You see, there are really many advantages to learning `vi`.

6.3.1 Immediate Takeoff

If you cannot wait to use `vi` and make your first text file: here you go. Type `vi` to start the editor. Press `i` to start the insertion mode and enter your text. You start a new line by hitting Enter. With Esc you leave the insertion mode.

With `:wq` ***filename*** you quit `vi` and save your text in a file called *filename* whereas with `:q!` you quit without saving. That's it. `vi` can be simple!

6.3.2 Starting vi

Now let us get serious. To start `vi`, enter: `vi` ***filename***, where filename is the name of the file you want to edit. If the file does not exist, `vi` will create it for you. You can also start `vi` without giving any filename. In this case, `vi` will ask for one when you quit or save your work. After you called `vi`, the screen clears and displays the content of the file *filename*. If it is a new file, it does not contain any text. Then `vi` uses the tilde character (~) to indicate lines on the screen beyond the end of the file. `vi` uses a cursor to indicate where your next command or text insertion will take effect. The cursor is the small rectangle, which is the size of one character, and the character inside the rectangle is called the current character. At the bottom of the window, `vi` maintains an announcement line, called the *mode line*. The mode line lists the current line of the file, the filename and its status. Let us now start `vi` with the new file *text.txt*. The screen should then look like Terminal 29. Please note that I have deleted some empty lines in order to save rain forest, i.e. paper. The cursor is represented by [].

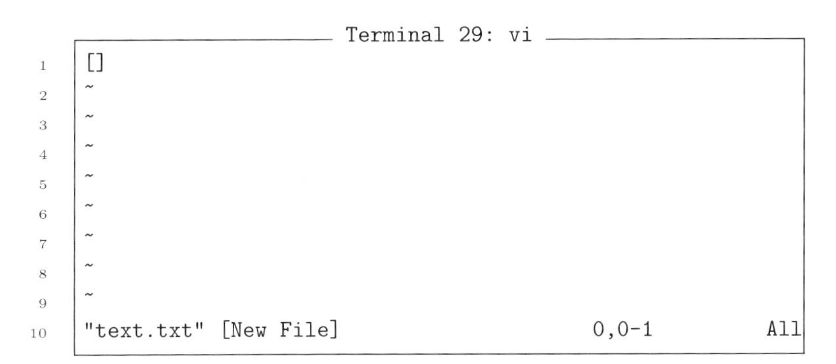

6.3.3 Modes

Line 10 in Terminal 29 shows the mode line. At this stage you cannot enter any text because `vi` runs currently in the command mode. In order to enter text, the input mode must be activated. To switch from the command mode to the input mode, press the `i` key (you do not need to press (Enter)). `vi` lets you insert text beginning at the current cursor location. To switch back to command mode, press (Esc). You can also use (Esc) to cancel an unfinished command in command mode. If you are uncertain about the current mode, you can press (Esc) a few times. When `vi` beeps, you have returned to the command mode. Okay, let us change to the input mode and enter some text.

Terminal 30: vi

```
1  This is new text in line 1. Now I press ENTER
2  and end up in the second line. I could also write to the e
3  nd of the line. The text will be wrapped automatically[]
4  ~
5  ~
6  ~
7  ~
8  ~
9  ~
10 -- insert --                            3,54          All
```

You can see some changes in the mode line in Terminal 30. "`-- insert --`" indicates that you are in the input mode. Furthermore, the current cursor position (line 3, column 54) is indicated. Now press (Esc) to get back into the command mode, "`-- insert --`" will disappear. Now let us save the file: press `:w`, and then (Enter). The mode line will display a message as shown in Terminal 31. If "`:w`" appears in your text you are still in the input mode!

Terminal 31: vi

```
1  This is new text in line 1. Now I press ENTER
2  and end up in the second line. I could also write to the e
3  nd of the line. The text will be wrapped automatically
4  ~
5  ~
6  ~
7  ~
8  ~
9  ~
10 "text.txt" [new] 3L, 160C written        3,54          All
```

Commands are very often preceded with the colon "`:`" character. Let us try another command: type `:set number`. Now you see line numbers in front of each line. Another command: type `:r !ls`. Woop. After hitting (Enter) you have a list of all files in your current working directory imported to your text file. That is magic, isn't it? `vi` run the shell command `ls` and imported the result into the text file at the current cursor position.

6.3.4 Moving the Cursor

Probably your screen is quite full now. Let us move the cursor. Usually you can use the arrow keys; but if they do not work you can use the following keys:

`h`	move one character to the left
`l`	move one character to the right
`k`	move up one line
`j`	move down one line

There are some more powerful commands for long-distance jumping.

`0`	move to beginning of the line
`$`	move to the end of the line
`H`	move to the top line of the screen
`M`	move to the middle line of the screen
`L`	move to the bottom line of the screen
`G`	move to the last line of the text
*n*G	move to the n-th line in the text
Ctrl+f	move one screen forward
Ctrl+b	move one screen backward

You have now learned some powerful tools for moving around in a file. You should memorize only the basic movements and know where to look up the others (you can download a cheat sheet at *www.kcomputing.com/vi.html*).

6.3.5 Doing Corrections

Maybe the most relaxing thing to know is that you can always *undo* changes by typing `u`. With `vim` you can even undo many commands whereas `vi` will recover only the last text change. If you are lucky, you can use the keys BkSp and Del in order to make *deletions* in the input mode. Otherwise you must make use of the commands in the command modus. To do so, first move the cursor so that it covers the first character of the group you want to delete, then type the desired command from the list below.

`x`	Delete only the current character
`D`	Delete to the end of the line
`db`	Delete from the current character to the beginning of the current word
`de`	Delete from the current character to the end of the current word
`dd`	Delete the current line
`dw`	Delete from the current character to the beginning of the next word

Notice that the second letter of the command specifies the same abbreviations as the cursor movement commands do. In fact, you can use delete with all of the cursor movement specifiers listed above, e.g. `dH` would delete everything from the current line to the top line off the screen.
In other cases you will need only to *replace* a single character or word, rather than deleting it. `vi` has change and replace functions, too. First move to the position where the change should begin (the desired line or the beginning of the desired word). Next, type the proper command from the list below. Finally, enter the correct text, usually concluded with Esc (except for `r`).

`cw`	Change a word.
`C`	Overwrite to the end of the line.
`r`	Replace a single character with another one. No Esc necessary.
`R`	Overwrite characters starting from the current cursor position.
`s`	Substitute one or more characters for a single character.
`S`	Substitute the current line with a new one.
`:r` *file*	Insert an external file at the current cursor position.

The change command `c` works like the delete command; you can use the text portion specifiers listed in the cursor movement list.

6.3.6 Save and Quit

`vi` provides several means of saving your changes. Besides saving your work before quitting, it is also a good idea to save your work periodically. Power failures or system crashes can cause you to lose work. From the command mode, you type `:w` (write) and hit Enter. In order to save the text in a new file, type `:w` ***`filename`***. You quit `vi` with `:q`. You can save and quit at once with `:x` or `:wq`. If you do not want to save your changes you must force quitting with `:q!`. Be cautious when abandoning `vi` in this manner because any changes you have made will be permanently lost.
Up to this point you have learned more than enough commands to use `vi` in a comfortable way. The next two sections explain some more advanced features which you might wish to use.

6.3.7 Copy and Paste

Frequently, you will need to cut or copy some text, and paste it elsewhere into your document. Things are easy if you can work with the mouse. When you mark some text with the mouse (holding the left mouse button) the marked text is in the memory (*buffer*). Pressing the right mouse button (or, on some systems, the left and right or the middle mouse buttons) pastes the text at the current cursor position. You can apply the same mechanism in the terminal window!
Things are bit more complicated if you have only the keyboard. First you cut or copy the text into temporary storage, then you paste it into a new location. Cutting means removing text from the document and storing it, while copying means placing a duplicate of the text in storage. Finally, pasting just puts the stored text in the desired location. `vi` uses a *buffer* to store the temporary text. There are nine numbered buffers in addition to an undo buffer. The undo buffer contains the most recent delete. Usually buffer 1 contains the most recent delete, buffer 2 the next most recent and so forth. Deletions

older than 9 disappear. However, vi also has 26 named buffers (a-z). These buffers are useful for storing blocks of text for later retrieval. The content of a buffer does not change until you put different text into it. Unless you change the contents of a named buffer, it holds its last text until you quit. vi does not save your buffers when you quit.
The simplest way to copy or move text is by entering the source *line numbers* and the destination line numbers. The m command moves (cuts and pastes) a range of text, and the t command transfers (copies and pastes) text. The commands have the syntax shown below:

:*x*m*y*	Move line number x below line number y.
:*x*,*y*m*z*	Move the lines between and including x and y below line z.
:*x*t*y*	Copy line x below line y.
:*x*,*y*t*z*	Copy lines between and including x and y below line z.

Another way is to use *markers*. You can mark lines with a letter from a to z. These markers behave like invisible bookmarks. To set a mark you use mx, with x being a letter from a to z. You can jump to a mark with 'x. The following list shows you how to apply bookmarks to copy or move text. Note: Bookmarks and line numbers can be mixed.

m*x*	Set a bookmark at the current line. x can be any letter from a-z.
'*x*	Jump to bookmark x.
:'*x*'*y*co'*z*	Copy lines between and including bookmarks x and y below bookmark z.
:'*x*'*y*m'*z*	Move lines between and including bookmarks x and y below bookmark z.
:'*x*'*y* w *file*	Write lines between and including bookmarks x and y into a file named *file*.

One last method uses the commands d (delete) or y (yank). With this method you can make use of different *buffers*. Go to the line you wish to copy or cut and press yy (yank) or dd (delete), respectively. Then move the cursor to the line behind which you want to insert the text and type p (paste). In order to copy a line into a buffer type "*x*yy, with x being a letter from a-z. You insert the buffer with "*x*p. To copy more than one line precede the command yy with the number of lines. 2yy copies 3 lines and 3yw copies 3 words. You see, vi is very flexible and you can combine many commands. If you are going to work a lot with it you should find out for yourself which commands you prefer.

6.3.8 Search and Replace

Finally, let us talk about another common issue: searching and replacing text. As files become longer, you may need assistance in locating a particular instance of text. `vi` has several search and search-and-replace features. `vi` can search the entire file for a given string of text. A string is a sequence of characters. `vi` searches forward with the slash (/) or backward with the question mark key (?). You execute the search by typing the command, then the string followed by (Enter). To cancel the search, press (Esc) instead of (Enter). You can search again by typing `n` (forward) or `N` (backward). Also, when `vi` reaches the end of the text, it continues searching from the beginning. This feature is called *wrapscan*. Of course, you can use wildcards or regular expressions in your search. We will learn more about this later in Section 9.3 on page 139. Let us take a look at the search-and-replace commands:

`/xyz`	Search forward for xyz.
`?xyz`	Search bachward for xyz.
`n`	Go to the next occurrence.
`N`	Go to the previous occurrence.
`:s/`*abc*`/`*xyz*`/`	Replace the first instance of *abc* with *xyz* in the current line.
`:s/`*abc*`/`*xyz*`/g`	Replace all instances of *abc* with *xyz* in the current line.
`:s/`*abc*`/`*xyz*`/gc`	Ask before replacing each instance of *abc* with *xyz* in the current line.
`:%s/`*abc*`/`*xyz*`/gc`	Ask before replacing each instance of *abc* with *xyz* in the whole file.
`:x,ys/`*abc*`/`*xyz*`/g`	Replace all instances of *abc* with *xyz* between lines *x* and *y*.

If you have followed this section about `vi` up to this stage, you should have obtained a very good overview of its capabilities. It can do more and it offers a whole range of options that one could set to personal preferences. However, since I do not believe that anyone is really going deeper into this, I stop at this point. You are welcome to read some more lines about `vi`. There are even whole books dedicated to its application [6, 9, 11].

Exercises

Now sit down and play around with some text. This is elementary!

6.1. Create a text file named *fruits.txt* using `cat`. Enter some fruits, one in each line. Append some fruits to this file.

6.2. Create a second text file named *vegetable* containing a list of vegetables, again, one item per line. Now concatenate fruits and vegetables onto the screen and into a file named *dinner*.

6.3. Sort the content of *dinner*.

6.4. Take some time and exercise with `vi`. Open a text document with `vi` or type some text and go through the description in this section. You must know the basic commands in order to write and edit text files!

7

Using the Shell

The power of Linux lies in its shell. Of course, it is nice and comfortable to run and control programs with a graphical user interface and the mouse. The maximum of flexibility, however, you gain from the shell. In this chapter you will learn some basic features of the shell environment.

7.1 What Is the Shell?

One of the early design requirements for Unix was the need for a small kernel, which is the memory-resistant part of the operating system. As already mentioned before, the kernel interfaces the user applications with the hardware (see Chap. 2 on page 9). In order to reduce the amount of code in the kernel, and thus the memory requirement of the kernel, many parts of the operating system were moved into user processes. By this, the directory */bin* filled up with commands like `ls`, `mv`, `mkdir` and so on. You access these commands via the shell. In fact, everything we have done so far and will do later happens "in the shell".

Since we run commands from the shell, the shell can be seen as a *command interpreter*. It recognizes the command and takes care of its correct execution. This involves handling the hardware via the kernel. In this section we will see that the shell also provides a powerful *programming language*. Thus, the shell is a command interpreter and a programming language. With the programming language you can easily automate processes or combine many commands to one. If you have some experience with DOS: the shell resembles the DOS file *command.com* and DOS's programming language for *batch files*. You will learn more about the power of the shell.

7.2 Different Shells

Until now we have always talked about *the* shell. However, there are *many* shells around. We will mainly work with the `bash` shell. I wrote *bash* like a command, because it is a command. Whenever you log into a system you can type `bash`. This opens a new bash shell for you. You can exit the shell with `exit`. When you type `exit` in your login shell (the shell you are in after you have logged in to the system) you actually logout! When you try to logout from a shell other than your login shell you will see an error message.

Terminal 32: Shells

```
1  login as: Freddy
2  Sent username "Freddy"
3  Freddy@192.168.1.5's password:
4  [Freddy@rware2 Freddy]$ bash
5  [Freddy@rware2 Freddy]$ logout
6  bash: logout: not login shell: use 'exit'
7  [Freddy@rware2 Freddy]$ exit
8  exit
9  [Freddy@rware2 Freddy]$ sh
10 sh-2.05b$ exit
11 exit
12 [Freddy@rware2 Freddy]$ ksh
13 -bash: ksh: command not found
14 [Freddy@rware2 Freddy]$ csh
15 [Freddy@rware2 ~]$ exit
16 exit
17 [Freddy@rware2 Freddy]$ echo $0
18 -bash
19 [Freddy@rware2 Freddy]$
20
```

Terminal 32 and Fig. 7.1 illustrate how different shell levels are opened. All these shells are called interactive shells because they wait for your input.

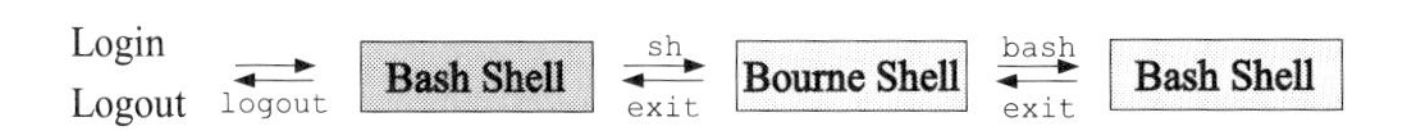

Fig. 7.1. From the login shell (*dark grey*) one can open new shells (*light grey*). In order to logout again one has to "go back" to the login shell by exiting all opened shells

In Terminal 32 the user Freddy logs in to the system and ends up in the login shell in line 4. Now, Freddy opens a new bash shell with the command `bash`. If he tries to logout from this shell, Freddy sees an error message (line 6). However, he can exit the shell with the `exit` command and gets back to

his login shell in line 9. From here, Freddy could logout with the command `logout`. Now you see the difference between `logout` and `exit`.
The bash shell is one of the modern shells. The name is an acronym, standing for *Bourne-Again Shell.* The first shell was the Bourne shell developed in 1978 by Steve Bourne from the Bell Laboratories. All other shells were developed later. Nowadays, you will find on most systems the *Bourne Shell* (`sh`), *C Shell* (`csh`) and *Bash Shell* (`bash`). You can see in Terminal 32 on the facing page how to enter these shells with their respective command. If a shell is not installed on the system, you will see an error message as in line 13. In line 12 Freddy tried to open the *Korn Shell* (`ksh`) but found that it is not installed on his system. Terminal 32 on the preceding page also displays differences of the *command line prompt* (that is something like "[`Freddy@rware2 Freddy`]$"). In all other terminals we omitted this part and you will just see the $ character (take a look at Sect. A.3.2 on page 267 for an example of how to change the shell prompt). Here, the bash prompt says that the user *Freddy* is logged in to the system *rware2* and is currently in his home directory called *Freddy.* In contrast, the Bourne shell prompt in line 10 shows only the version of the shell, and the C shell prompt in line 15 looks similar to the bash prompt. All these prompts can be set up according to your personal preferences. If you do not know in which shell you are, type `echo $0` as shown in line 17 of Terminal 32 on the preceding page. This command is available in all shells.

7.3 Setting the Default Shell

When you login to your system you can use the command `echo $0` to check out your default shell. You can easily change your default shell with the command `chsh` (change shell).

Terminal 33: chsh

```
1  $ cat /etc/shells
2  /bin/sh
3  /bin/bash
4  /sbin/nologin
5  /bin/bash2
6  /bin/ash
7  /bin/bsh
8  /bin/tcsh
9  /bin/csh
10 $ chsh
11 Change Shell for Freddy.
12 Password:
13 New Shell [/bin/bash]: /bin/csh
14 Shell changed.
15 $
```

In Terminal 33 on the preceding page we see how to find out which shells are installed on your system. There is a list available in the file */etc/shells*. On my system, I can choose between 8 different shells. In order to change the login shell, type `chsh`. Now you have to enter your password and give the path to your desired shell. The new shell will be activated after a new login.

7.4 Useful Shortcuts

The bash shell provides some very nice shortcuts. These make your life easier when working in the terminal. One very nice feature is the history file. The bash shell remembers all the commands you typed. They are saved in a file in your home directory. The file is called *.bash_history*. You can scroll through your last commands with the arrow keys (↑ and ↓). With Strg+R you can search the history file for a command. After you have typed Strg+R you see the message "reverse-i-search:". If you now type characters into the command line, bash is querying the *.bash_history* file for the most recent match. With Enter you can execute the command and by pressing Tab you paste the command into the command line and you can edit it. If you instead hit Strg+R again, bash searches for the next match in the history file. Try it out to get a feeling for it; but try it with commands like `cd`, not with commands like `rm`, otherwise you might, by error, delete a file.
There is an alternative way to search the history file: by using `grep`. This is very helpful when you remember that you have typed a certain command in the past few days but you do not want to use the up arrow 400 times to find it. The next time you find yourself in such a situation, type:

`history|grep -i` *`keyword`*

This will read the history file and perform a case-insensitive search on the keyword you are looking for.

Terminal 34: Query Command History

```
1  $ history|grep -i sort
2    140  man sort
3    141  info sort
4    143  info sort
5    816  sort amino1>amino1s; sort amino2>amino2s
6   1060  ls -l |sort +4
7   1066  ls -l |sort +4 -nr
8   1077  alias llss="lls|sort +4 -n"
9   1085  history|grep -i sort
10 $ !1060
11 ls -l |sort +4 -nr
12 -rwxrw-r-- 1 Freddy Freddy 321 Mai 17 11:54 spaces.sh
13 -rw-rw-r-- 1 Freddy Freddy  59 Mai 17 12:40 err.txt
14 -rw-rw-r-- 1 Freddy Freddy  11 Mai 17 13:51 list.txt
```

```
15  -rw-rw-r-- 1 Freddy Freddy   3 Mai 17 13:58 new
16  insgesamt 20
17  $
```

Terminal 34 demonstrates its use. In line 1 we search for all past commands in the history file that used the command `sort`. There are 8 hits, the most recent one being at the bottom in line 9. Each command is preceded by an unambiguous identifier. We can use this identifier to execute the command. This is done in line 10. Note that you need to precede with an exclamation mark (`!`). This is cool, isn't it?
The following list shows you a couple of shortcuts that work in the bash shell:

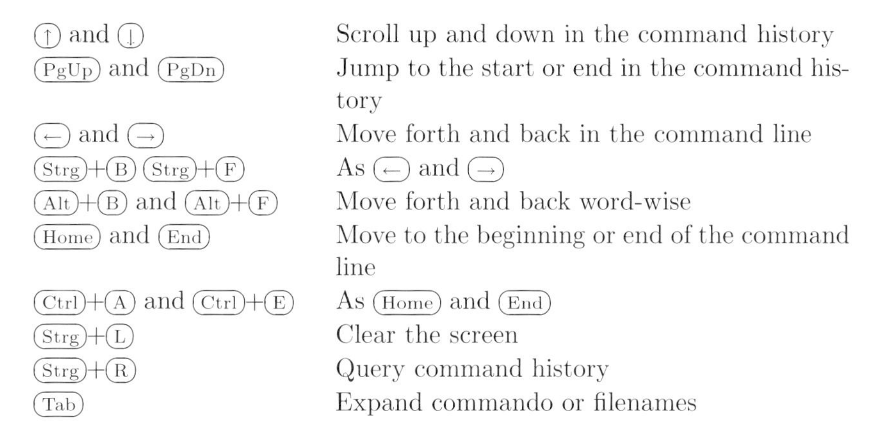

(↑) and (↓)	Scroll up and down in the command history
(PgUp) and (PgDn)	Jump to the start or end in the command history
(←) and (→)	Move forth and back in the command line
(Strg)+(B) (Strg)+(F)	As (←) and (→)
(Alt)+(B) and (Alt)+(F)	Move forth and back word-wise
(Home) and (End)	Move to the beginning or end of the command line
(Ctrl)+(A) and (Ctrl)+(E)	As (Home) and (End)
(Strg)+(L)	Clear the screen
(Strg)+(R)	Query command history
(Tab)	Expand commando or filenames

The last shortcut in this list deserves our special attention. After you have entered the first characters of a command, you can press the tabulator key (Tab). If the command is already unambiguously described by the characters you entered, the missing characters will be attached automatically. Otherwise you hear a system beep. When you press the tabulator key twice, all possible command names will be displayed. The same functions with filenames. Try it out! This is really a comfortable feature!

7.5 Redirections

We have already used redirections in previous sections. The main purpose was to save the output of a command into a file. Now let us take a closer look at redirections. In Unix/Linux, there are three so-called *file descriptors*: standard input (stdin), standard output (stdout) and standard error (stderr). By default, all programs and commands read their input data from the standard input. This is usually the keyboard. When you enter text using the text editor `vi`, the text comes from the keyboard. The data output of all programs is sent to the standard output, which by default is the screen. When you use the command `ls`, the result is printed onto the screen (or the active terminal

when you use X-Windows). The standard error is displayed on the screen, too. If you try to execute the non-existing command `lss`, you will see an error message on the screen.
It is often convenient to be able to handle error messages and standard output separately. If you do not do anything special, programs will read standard input from your keyboard, and they will send standard output and standard error to your terminal's display. The shell allows you to redirect the standard input, output and error. Basically, you can redirect stdout to a file, stderr to a file, stdout to stderr, stderr to stdout, stderr and stdout to a file, stderr and stdout to stdout, stderr and stdout to stderr.
As already mentioned above, standard input normally comes from your keyboard. Many programs ignore stdin. Instead, you enter e.g. filenames together with the command. For instance, the command `cat filename` never reads the standard input; it reads the filename directly. However, if no filename is given together with the command, Unix/Linux commands read the required input from stdin. Do you remember? We took advantage of this trick in Section 6.1 on page 64. In Table 7.1 you find a summary of the syntax for redirections. You should be aware that different shells use different syntax!

Table 7.1. Common Standard Input/Output Redirections for the C Shell and Bourne Shell. The numbers "1" and "2" stand for the standard output and the standard error, respectively

Function	**csh**	**sh**
Send stdout to file	prog > file	prog > file
Send stderr to file		prog 2> file
Send stdout and stderr to file	prog >& file	prog > file 2>&1
Take stdin from file	prog < file	prog < file
Append stdout to end of file	prog >> file	prog >> file
Append stderr to end of file		prog 2>> file
Append stdout and stderr to end of file	prog >>& file	prog >> file 2>&1

Now let us take a look at some examples from the `bash` shell. In the following list, the numbers "1" and "2" stand for the standard output and the standard error, respectively. Keep this in mind. This will help you in understanding the following examples.

`ls -l > list.txt`	Output is written into the file *list.txt*
`ls -l 1> list.txt`	Output is written into a file *list.txt*
`ls -l >> list.txt`	Output is appended to the file *list.txt*
`ls -e 2> error.txt`	Error is written into the file *error.txt*
`ls -e 2>> error.txt`	Error is appended to the file *error.txt*
`ls >> all.txt 2>&1`	Output and error are appended to the file *all.txt*
`ls &> /dev/null`	Send all output to the nirvana. The program will execute its task quietly. `/dev/null` is the nirvana, a special device that disposes of unwanted output
`cmd < file1 > file2`	The hypothetical command `cmd` reads *file1* and sends the output to *file2*

You see, there are many ways to redirect the input and output of programs. Let us finally consider that you wish to redirect the output of a command into a file *and* to display it on the screen. Do you have to run the command twice? Not with Unix and Linux! You can use the command `tee`. With

```
ls | tee filename
```

you see the content of the current directory on the screen and save it in a file named *filename*. With the option `-a`, the result is *appended* to the file *filename*.

7.6 Pipes

Pipes let you use the output of one program as the input of another one. For example you can combine

```
ls -l | sort
```

to sort the content of the current directory. We will use pipes later in conjunction with `sed`. For example, in

```
ls -l | sed s/txt/text/g
```

the command `ls -l` is executed and sends (pipes) its output the `sed` program that substitutes all occurrences of *txt* by *text*. Another common way to use a pipe is in connection with the `grep` command (see Sect. 6.1.5 on page 68). With the line

```
cat publications | grep AIDS
```

all lines of the file called *publications* that contain the word *AIDS* will be displayed. Another great pipe you should remember is the combination with the command `less` (see Sect. 6.1.3 on page 67). This command displays large texts in a rather nice way: you can scroll through the output. You get back to the command line after hitting Q. Imagine you have a large number of files in your home directory and run `ls -l`. Most probably, the output will not fit onto the screen. This means that you miss the beginning of the text. In such cases you should use

```
ls -l | less
```

in order to be able to scroll through all the output. The following command offers an interesting combination and a good example of the power of redirections and pipes:

```
ls -l | tee dir-content | sort +4 -n > dir-content-size-sorted
```

The file *dir-content* will contain the content of the current directory. The screen output is piped to `sort` and sorted according to file size (the option `+4` points to column 4). The result is saved in the file *dir-content-size-sorted.*

7.7 Lists of Commands

Usually, if you want to execute several commands in a row, you would write a shell script. However, sometimes it is faster to write the commands on the command line. A list of commands is a sequence of one or more commands separated by ";", "&&", or "||". Commands separated by a ";" are executed sequentially; the shell waits for each command to terminate. The control operators "&&" and "||" denote *AND* lists and *OR* lists, respectively. An AND list has the form

```
command1 && command2
```

`Command2` is executed if, and only if, `command1` returns an exit status of zero (this means that no error has occurred). An OR list has the form

```
command1 || command2
```

`Command2` is executed if, and only if, `command1` returns a non-zero exit status (that is an error). Let us take a look at one example.

Terminal 35: Command List

```
1  $ find ~ -name "seq" && date
2  /home/Freddy/seq
3  /home/Freddy/neu/seq
4  /home/Freddy/neu/temp/seq
5  Mit Mai 28 21:49:25 CEST 2003
6  $ find ~ -name "seq" || date
7  /home/Freddy/seq
8  /home/Freddy/neu/seq
9  /home/Freddy/neu/temp/seq
10 $
```

In both cases in Terminal 35 the command `find` finishes execution without error. This leads to the execution of the `date` command in the first (`&&`), but not in the second case (`||`).

7.8 Aliases

A very nice feature is the alias function. It allows you to create shortcuts (aliases) for every command or combination of commands. Let us assume you have pretty full directories. The command `ls -l` fills up the screen and you even lose a lot of lines because the screen can display only a limited number of lines. Thus, you wish to pipe the output of `ls -l` to the command `less`. Instead of typing every time `ls -l | less` you can create an alias. The alias is created with the command `alias`:

```
alias lls="ls -l | less"
```

When you now type `lls` the command `ls -l | less` is executed. You can also use your alias in a new alias:

```
alias llss="lls | sort +4 -nr"
```

sorts the output according to file sizes, beginning with the largest file. If you want to see a list of all your active aliases type `alias`. To remove an alias use the command `unalias`, like `unalias llss`.
Aliases are not saved permanently if you do not explicitly tell the system to do so. Furthermore, they are valid only in the active shell. If you open a new shell, you have no access to your aliases. In order to save your alias you must place an entry into the file *.bashrc* in your home directory (use the editor `vi`). Since different systems behave differently, you are on the safe side by putting the same entry into the *.bash_profile* in your home directory (see Sect. A.3.1 on page 266).
Much more powerful than aliases are shell scripts. With shell scripts you can also use parameters and perform much more sophisticated tasks. You will learn more about shell scripts in Chapter 8 on page 97.

7.9 Scheduling Commands

Unix/Linux gives you the possibility to schedule commands or scripts. It works like an alarm clock. You set the time and date when the alarm should ring. Instead of ringing a bell, a command will be executed. One possible example is to instruct Linux to update a database every night. Or how about writing a script that removes all old backup files every evening?
When you start the Unix/Linux system, a number of background processes are started. You can see this when you follow the lines on the startup screen.

These background system programs are called *daemons*. Among these daemons is *cron*. Cron checks up a system table for entries. These entries tell cron when to do what. Note: On most systems, you must get permission from the system administrator before you can submit job requests to cron. In order to fill the table with commands that need to be executed repeatedly (e.g. hourly, daily or weekly), there is a program called `crontab`. The `crontab` command creates a crontab file containing commands and instructions for the cron daemon to execute. For editing the crontab file `crontab` uses the text editor `vi` (see Sect. 6.3 on page 72).
You can use the `crontab` command with a number of options. The most important ones are:

crontab -e	Edit your crontab file. If not existent, create the crontab file.
crontab -l	Display your crontab file.
crontab -r	Remove your crontab file.

Each entry in a crontab file consists of six fields, specifying in the following order: minute, hour, day, month, weekday and command(s). The fields are separated by spaces or tabs. The first five fields are integer patterns and the sixth is the command to execute. Table 7.2 briefly describes each of the fields.

Table 7.2. Crontab File Field Entries

Field	Value	Description
minute	0-59	The exact minute when the command is to be executed
hour	0-23	The hour of the day when the command is to be executed
day	1-31	The day of the month when the command is to be executed
month	1-12	The month of the year when the command is to be executed
weekday	0-6	The day of the week when the command is to be executed (Sunday = 0, Monday = 1, Tuesday = 2 and so forth)
command		The complete sequence of commands to execute. Commands, executables (such as scripts), or combinations are acceptable

Each of the values from the first five fields in Table 7.2 may be either an asterisk (*), meaning all legal values, or a list of elements separated by commas. An element is either a number or an inclusive range, indicated by two numbers separated by a dash (e.g. 10–12). You can specify days with two fields: day of the month and day of the week. If you specify both of them as a list of elements, cron will observe both of them. For example

```
0 0 1,15 * 1 /mydir/myprogram
```

would run the program *myprogram* in the *mydir* directory at midnight on the 1st and 15th of each month and on every Monday. To specify days by only one field, the other field should be set to *. For example, with

```
0 0 * * 1 /mydir/myprogram
```

the program would run only on Mondays at midnight.
Crontab can also be used with lists, ranges and steps. A list is a set of numbers (or ranges) separated by commas. Examples are `1,2,5,9` or `0-4,8-12`. Ranges of numbers are two numbers separated by a dash. The specified range is inclusive. For example, `8-11` for an hour entry specifies execution at hours 8, 9, 10 and 11. Step values can be used in conjunction with ranges. Following a range with `/5` specifies 5 skips through the range (every fifth). For example, `0-23/2` can be used in the hour field to specify command execution every other hour. This is much less typing then `0,2,4,6,8,10,12,14,16,18,20,22`. Steps are also permitted after an asterisk, so if you want to execute a command every 2 hours, just use `*/2`.

7.10 Wildcards

You do not have to be a programmer to appreciate the value of the shell's *wildcards*. These wildcards make it much easier to search for files or content. When the shell expands a wildcard, it is replaced with all the matching possibilities. Let us take a look at the most common wildcards:

- `*` Matches anything or nothing. "`protein*`" matches *protein*, *proteins*, *protein-sequence* and so on.
- `?` Matches any single character. For example, "`seq?.txt`" matches *seq1.txt* but not *seq.txt*.
- `[ ]` Matches any character or character range included. "[`A-Z`]*" matches anything starting with an uppercase character. "[`sf`]" matches the characters *s* and *f*.
- `[! ]` Inverses the match of [ ]. "[`!a-zA-Z`]" matches anything but letters.

These are the most important wildcards. Let us see how we can use them in conjunction with the command `ls`.

Terminal 36: Wildcards

```
1  $ ls
2  1file  cat      list.sh   new       seq2.txt  spaces.sh
3  AFILE  err.txt  list.txt  seq1.txt  seq5.txt  end.txt
4  $ ls [n]*
5  new
```

```
6  $ ls seq[0-4].txt
7  seq1.txt  seq2.txt
8  $ ls [A-Z]*
9  AFILE
10 $ ls *[!a-z]*
11 1file  err.txt  list.txt  seq2.txt  spaces.sh
12 AFILE  list.sh  seq1.txt  seq5.txt  end.txt
13 $
```

Terminal 36 shows you some examples of the application of wildcards in connection with the `ls` command. In line 1 we obtain an overview of all files present in the working directory. In line 4 we list only files that begin with the lowercase character *n*. In line 6 we restrict the output of the `ls` command to all files that begin with "seq", followed by a number between 1 and 4, and ending with ".txt". The `ls` command in line 8 resembles line 4; however, the files must begin with any uppercase character. Finally, in line 10, we list all files with filenames that contain a character other than a lowercase character. Of course, you can use wildcards also with commands like `rm`, `cp` or `mv`. However, it is always a good idea first to check your file selection with `ls`. You might have an error in your pattern and thus erase the wrong files.

7.11 Processes

In Linux, a running program or command is called a process. Ultimately, everything that requires the function of the processor is a process. The shell itself is a program, too, and thus a process. Whenever one program starts another program, a new process is initiated. We are then talking about the *parent process* and the *child process*, respectively. In the case that a command is executed from the shell, the parent process is the shell, whereas the child process is the command you execute from the shell. That in turn means that, when you execute a command, there are two processes running in parallel: the parent process and the child process. As we heard before in Chapter 2 on page 9, this is exactly one strength of Unix/Linux. The capacity of an operating system, such as Unix or Linux, to run processes in parallel is called *multitasking*. In fact, with Unix or Linux you can run dozens of processes in parallel. The operating system takes care of distributing the calculation capacity of the processor to the processes. This is like having a nice toy in a family with several children. There is only one toy (processor). The mother (operating system) has to take care that every kid (process) can play with the toy for a certain time before it is passed to the next child. In a computer this "passing the toy" takes place within milliseconds.
As a consequence of multitasking, it appears to the user as if all processes run in parallel. Even though you might run only a few processes, the computer might run hundreds of processes in the background. Many processes are running that belong to the operating system itself. In addition, there might be

other users logged in to the system you are working on. In order to unambiguously name a process, they are numbered by the operating system. Each process has its own unique *process number.*

7.11.1 Checking Processes

To get an idea about the running processes you can use the command ps.

Terminal 37: List Processes

```
1  $ ps
2    PID TTY          TIME CMD
3    958 pts/0    00:00:00 bash
4   1072 pts/0    00:00:00 ps
5  $ ps -T
6    PID TTY      STAT   TIME COMMAND
7    958 pts/0    S      0:00 -bash
8   1073 pts/0    R      0:00 ps -T
9  $
```

Terminal 37, lines 1 to 4 show the output of the ps command (print process status). The first line is the header describing what the row entries mean:

PID	Process ID	The process ID (identification) is a unique number assigned to each process by the operating system.
TTY	Terminal	The terminal name, from which the process was started.
STAT	State	This entry indicates the state of the process. The process can either sleep (S), run on the processor (O), wait to run on the processor (R), just be started (I), paused (T) or be "zombie" (Z), that is ending.
TIME	Run Time	The run time corresponds to the amount of processor computing time the process has used.
CMD	Command	This is the name of the process (program or command name).

As usual, there are a lot more options possible. Quite useful is the option -T. It shows you all processes of the terminal in which you are currently working together with the process state. However, the most important information for you are the process ID and name. In the example in Terminal 37 there are two processes running. The first process is the bash shell and the second process the command ps itself. They have the process IDs 958 and 1072, respectively. The output of ps -T is shown in lines 6 to 8. Of course, the shell is still running together with ps -T. The *process's state* information can be interesting to identify a stopped (halted) process.

7.11.2 Background Processes

There are two basic ways of running processes: either in the *foreground* or in the *background*. In foreground mode, the parent process (usually the shell) is suspended while the child (the command or script) process is running and taking over the keyboard and monitor. After the child process has terminated, the parent process resumes from where it was suspended.
In background mode, the parent process continues to run using the keyboard and monitor as before, while the child process runs in the background. In this case it is advisable that the child process gets all its input from and sends all its output to files, instead of the keyboard and monitor. Otherwise, it might lead to confusion with the parent process's input and output. When the child process terminates, either normally or by user intervention, the event has no effect on the parent process, though the user is informed by a message sent to the display.
An example would be the command `sort`. For long text files execution of `sort` takes very long. Thus, you might want to start it directly in the background. That is done as shown in Terminal 38.

Terminal 38: Background Program Execution

```
1  $ sort largefile.txt > result.txt &
2  [3] 23001
3  ...
4  $ [3]- Done  sort largefile.txt >result.txt
5  $
```

Executing any command with the ampersand (`&`) will start that command in the background. The `sort` command is started in the background as shown in line 1 of Terminal 38. You will then see a message giving the *background ID* (`[3]`) and the *process ID* (`23001`). (On many systems you have to press (Enter) before you see the message. This can be changed for the current session by typing `set -b`). Now you can continue to work in the shell. When the background process is finished you will see a message as shown in line 4.
You may also elect to place a process already running in the foreground into the background and resume working in the shell. This is done by pressing (Ctrl)+(Z) and then typing `bg`. This brings the process into the background. In order to bring a background process back into the foreground, you have to type `fg processname` or `fg backgroundID`. You can again send it back into the background by stopping (not quitting) the process with (Ctrl)+(Z) and then typing `bg`. Note that you quit a process with (Ctrl)+(C).

Terminal 39: Background Processes

```
1  $ ps -x
2    PID TTY      STAT   TIME COMMAND
3   1050 pts/1    S      0:00 -bash
4  23036 pts/1    R      0:00 ps -x
5  $ sleep 90 &
```

```
 6  [1] 23037
 7  $ ps -x
 8    PID TTY      STAT   TIME COMMAND
 9   1050 pts/1    S      0:00 -bash
10  23037 pts/1    S      0:00 sleep 90
11  23038 pts/1    R      0:00 ps -x
12  $ fg sleep
13  sleep 90
14
15  [1]+  Stopped                 sleep 90
16  $ ps -x
17    PID TTY      STAT   TIME COMMAND
18   1050 pts/1    S      0:00 -bash
19  23037 pts/1    T      0:00 sleep 90
20  23039 pts/1    R      0:00 ps -x
21  $ bg
22  [1]+ sleep 90 &
23  $
```

A good command to play around with is `sleep 90`. This command just waits 90 seconds. Of course, you can use any other desired amount of seconds. Terminal 39 shows an example. In line 1 we list all active processes. In line 6, we execute `sleep 90` as a background process. In line 12 we bring `sleep` to the foreground with the command `fg`. Then, which is not visible in Terminal 39, we press (Ctrl)+(Z), thereby pausing (stopping) the process. With the `bg` command we resume execution of `sleep` in line 21.
You may actively terminate (kill) any process, provided you own it (you own any process that you initiate plus all descendants (children) of that process). You can also downgrade (but not upgrade!) the priority of any owned process.

7.11.3 Killing Processes

Sometimes a process, like a program, *hangs*. This means you cannot access it any more and it does not react to any key strokes. In that case you can *kill* the process. This is done with the command `kill PID`, where PID is the process ID you identify with `ps`. When you kill a process, you also kill all its child processes. In order to kill a process you must either own it or be the superuser of the system.

Terminal 40: Process Killing

```
1  $ sleep 60 &
2  [1] 1009
3  $ ps
4    PID TTY          TIME CMD
5    958 pts/0    00:00:00 bash
6   1009 pts/0    00:00:00 sleep
7   1010 pts/0    00:00:00 ps
```

```
8  $ kill 1009
9  [1]+  Terminated  sleep 60
10 $ ps
11   PID TTY          TIME CMD
12   958 pts/0    00:00:00 bash
13  1011 pts/0    00:00:00 ps
14 $
```

Terminal 40 illustrates process killing. In line 1 we start the process `sleep 60`. With the `ps` command we find out that the process's ID is 1009. In line 8 we kill the process with `kill 1009`. We get a message that the process has been killed. You might see that message only after hitting (Enter) once.
Some processes can be really hard to kill. If the normal `kill` command does not help, try it with option `-9`. Thus the command becomes `kill -9 PID`, with PID being the process ID. This should do the job.

Exercises

If you have read this chapter carefully, you should have no problems with the two little exercises...

7.1. Set `bash` as your default shell.

7.2. Print out a list of the active processes, sort the list by command names and save it into a file.

8 Shell Programming

With this chapter we enter a new world. Until now you have learned the basics of Unix/Linux. You have learned how to work with files and create and edit text files. Now we are going to *use* Linux. We take advantage of its power. Up to now everything was quite uncomfortable and I guess that you thought occasionally: "Okay, it is free – but damn uncomfortable!" But now you are a pro! You know what it is about, how to work on the system; now it is time to take advantage of it, squeeze it out, form it, make it working for you, harvest the fruits of learning – by more learning. In this section you start programming! If you thought programming is something for the freaks – forget it. Everybody can do it, but you must like solving problems. Hey, you are a scientist! That is your possession! And it is creative. It is like art – you will see! It is like solving crosswords: you need to take your time and passion and probably need to look up one or the other thing; but finally you solved it.
Throughout the last chapters we used the shell intensively. Now it is time to take a closer look at its functions.

8.1 Script Files

When we write a shell program we will save the program code in a file. We call this file a *script file* and the program itself a *script*. Script files should be named with relatively short names that describe their function. Otherwise you will end up in chaos. It is more than a good idea to give script files the extension *.sh*. This is crucial to keep something called recursion from occurring. Recursion is a kind of futile cycle, something that repeatedly executes itself. For example, consider the filename *date* for a script file. This sounds like a clever name for a file that does something with dates. However, as you know, `date` is also a command. If you now happen to call the command `date` from your script called *date* you start a futile cycle. Every time the shell interprets the line in the script file containing *date* a new instance of your script will be

started. Now your script is running twice; but the creation of new instances goes on and on and on until your system is completely busy with *date* and gives an error message. Most probably you have to restart your system (or the system administrator will have to do this). Always use file extensions for script files.
What does the inside of a script file look like? Well, it is simply a file containing commands to be executed. In this section we will write shell script files; however, later we will write `sed`, `awk` and `perl` script files. Remember: `sed`, `awk` and `perl` are script languages.
In the first line of a shell script file you identify the program that is to interpret the script. For example, in order to invoke the bash shell to interpret the script, the first line would look like `#!/bin/bash`. This line consists of the so-called *shebang* "`#!`" (from *sharp* and *bang*). In the next lines the script follows. You can and should make use of the possibility to comment your program. Otherwise, you will soon forget what your script does and why. Ergo: always use comments! Comments begin with a hash (`#`). Everything after the hash is ignored by the shell.
Okay, let us write our first script. Script files must be executable. I guess this is clear?! A script is a program – and programs have to be executed. Our first script (Program 2) will convert all files ending with *.sh* in the home directory and all subdirectories into executable files.

Program 2: Create executable .sh files

```
1 #!/bin/bash
2 # save as con-sh-exe.sh
3 # This script converts .sh files to executables
4 echo "Search for *.sh files"
5 find ~ -name "*.sh"
6 echo "Perform task"
7 find ~ -name "*.sh" -exec chmod u+x {} \;
8 echo "Ready:"
9 find ~ -name "*.sh" -exec ls -l {} \;
```

Using the text editor `vi`, enter Program 2 and save it in a file called *con-sh-exe.sh*. Then make the file executable with

```
chmod u+x con-sh-exe.sh
```

In order to run the script you must type

```
./con-sh-exe.sh
```

Now, let us go step by step through the program. Line 1 instructs the bash shell to execute all the commands. In principle, we could also instruct another shell or any other command interpreter like `awk` or `perl`. Lines 2 and 3 contain a comment on the program. As said before, lines beginning with a hash (`#`) are ignored by the command interpreter. The command `echo` in lines 4, 6 and

8 prints out a message to the standard output (stdout), i.e. the screen. The message is placed between quotation marks (“”). You could also use `echo` without anything. That would print a blank line. In line 5 the script runs the command `find`. It searches in the home directory (the shortcut is ~, you could also write */home/Freddy*) for files having the extension *.sh*. Note that the filename is enclosed in quotation marks. The output of the command is printed, as usual, to the stdout. There is no difference whether you run the command from the command line or a script. In line 7 the command `find` is used again. Here the option `-exec` prevents the output to stdout. Instead, the files found are directed to the program executed by the option `-exec`. The files are provided in the form of empty curled brackets ({ }). Here, the program executed is `chmod` with the option `u+x` (see Sect. 4.4 on page 41). As said, the curled brackets represent the files found by `find` and for which the permission is to be changed by `chmod`. Note that the command is followed by “`\;`”. This is obligatory. In line 9 the `find` command is used to list all the files ending with *.sh*. This is done in order to visualize that the permissions are set correctly. If it comes down to the main task: change the permission of all files ending with *.sh* to executable; the script could be minimized to lines 1 and 7. All the rest is luxury but helps to understand what is going on.

8.2 Modifying the Path

In the example above we executed the script with `./con-sh-exe.sh`. We always have to supply the path to the script. If we entered `con-sh-exe.sh` we would receive the error message: “command not found”. However, we could add the directory where we save scripts (let us say we save them in *scripts*) and other executables to the system path. This is a variable that contains all directories that might contain commands. To see the actual state of the path variable enter `echo $PATH`.

Terminal 41: The Path Variable

```
1 $ echo $PATH
2 /usr/local/bin:/bin:/usr/bin:/usr/X11R6/bin
3 $
```

As shown in Terminal 41, the system checks by default the directories */usr/local/bin*, */bin*, */usr/bin* and *usr/X11R6/bin* for executable commands. If you wish to add the directory *scripts* to the path, you have to extend the content of the *PATH* variable. This can be done in the file *.bash_profile* or *.profile* (see Sect. A.3.1 on page 266). Open one of the files with the editor `vi` and add the following line:

```
PATH=$PATH:$HOME/scripts
```

The next time you start the bash shell you can execute scripts sitting in the directory *scripts* directly by entering their name. Alternatively, you can directly activate the new path by typing

```
. ../.bash_profile
```

Another option is to write a script that adds the current working directory to the *PATH* variable. Program 3 shows such a script.

Program 3: Add to Path

```
#!\bin\bash
# save as add2path.sh
# add the currently active directory to the path variable
# assignment only active in the current session
PATH=$PATH":"$(pwd)
```

This is a pretty short script. It actually assigns to the system variable *PATH* its original value, plus a colon (`:`) plus the directory you are currently in (`$(pwd)`). Note that `$( )` returns the result of the command enclosed in the parentheses, here, `pwd`.

8.3 Variables

A variable is a symbol that stands for some value – an abstraction for something more concrete. The shell's ordinary variables are very simple. A variable comes into existence when a value is assigned to it. Its value is what programmers call a *string*: simple text (it can be composed of numeric digits that would represent a number to the human observer or to some other program, but the shell itself is blind to their numeric value).
The assignment of a variable is a matter of only one line:

```
plant=orchid
```

There must be no spaces around the equal character. With this simple command the value "orchid" is assigned to the variable named *plant*. If you wish to recall the content saved in *plant*, you call it with the `echo` command:

```
echo $plant
```

Note that you have to precede the variable name with the dollar character (`$`).

Terminal 42: Variables

```
$ plant=orchid
$ echo $plant
orchid
$ orchid=parasite
$ plant=$orchid
$ echo $plant
parasite
```

```
8  $ echo "The value of \$orchid is $orchid"
9  The value of $orchid is parasite
10 $ echo "$animal"
11
12 $
```

Terminal 42 explains how to assign variables. In line 1 we create the variable *plant* and assign the value (content) "orchid" to it. With `echo $plant` we display the content of the variable. In line 4 we save the value "parasite" in the variable *orchid*. Then we assign to the variable *plant* the content of the variable *orchid*. Note that we overwrite the old content of *plant*. As you see in line 5, you can indirectly assign a value to a variable. Finally, in line 8 we use the `echo` command to print some text and the value of the variable *orchid*. Note that you must use the backslash (escape character) in order to print the special character `$`. Otherwise, the dollar character would indicate to the shell that the value of a variable should be printed. In line 10 we recall the content of the variable *animal*. This variable has not been created yet and thus is empty. After a variable has been declared, it is available only in the active shell and only during the active session. If you want to export a variable to other sessions you need to apply the command `export`. This is shown in Terminal 43.

Terminal 43: Export Variables

```
1  $ plant=rose
2  $ sh
3  sh$ echo $plant
4
5  sh$ exit
6  exit
7  $ export plant
8  $ sh
9  sh$ echo $plant
10 rose
11 sh$ exit
12 exit
13 $
```

In line 1 in Terminal 43 we create the variable *plant* and assign the value "rose" to it. Then we change into the Bourne shell with the command `sh`. Here, as depicted in line 3, we recall the value of *plant*. It is empty. Now we exit the Bourne shell, export the variable *plant* with `export plant` and go back into the Bourne shell.

Apart from the variables we create, there are a number of system variables we can use. These variables are also called *environmental variables* or *shell variables*. Environmental variables are available for the whole system and all shells. Shell variables are available only in the current shell. In Terminal 43 we exported a shell variable to the environment. With the command `env` (environment) you can list all available environmental variables. With the `set` command you can list all shell variables. Some interesting environmental variables are shown in Table 8.1.

Table 8.1. Some commonly used shell variables

Variable	Content	Value
SHELL	active shell	"/bin/bash"
USER	the user	"Freddy"
PATH	search path	"/bin:/usr/bin:/home/Freddy/scripts"
PWD	active directory	"/home/Freddy/scripts"
HOME	home directory	"/home/Freddy"

The list shows the variable name (note that system variables are always uppercase), its meaning and an example of its content.
In order to delete a variable, the command `unset` can be used. Then the variable is totally gone and not just empty.

Terminal 44: Variables

```
1  $ my_var=value
2  $ set|grep my_var
3  my_var=value
4  $ env|grep my_var
5  $ export my_var
6  $ env|grep my_var
7  my_var=value
8  $ unset my_var
9  $ env|grep my_var
10 $ set|grep my_var
11 $
```

The example shown in Terminal 44 shows how you can easily search for a variable in the shell or environment with the help of `grep`. In line 8 the variable *my_var* is removed from the system.

8.4 Input and Output

Of course, a script should be able to produce output and get some input. The shell scripting language provides different possibilities to do so. With the `echo` command you can display messages. With the commands `read` and `line` you can ask the user to provide some input. Finally, shell scripts can be directly fed with input when called from the command line – these are script parameters. In this section we will see examples of all of these.

8.4.1 echo

As we saw before, the echo command prints out text or the value of variables onto the screen. You should always enclose the text in double quotes, as shown in Terminal 42 on page 101 line 8. echo offers the interesting options -e, which makes echo evaluating the character behind a backslash:

\\	Backslash	\a	Alert, System Bell	\b	Backspace
\n	New Line	\t	Tabulator		

When you type echo -e "\a" you will hear the system bell. This function is quite nice to inform the user that the execution of a script has been finished. In a similar manner you can introduce tabulators or a new line. Thus, it is possible to format the output a little bit.

8.4.2 Here Documents: <<

Assume you want to write a couple of help lines in your script. Of course, you could achieve this with the echo command. However, it looks quite messy. For such cases the shell offers the << operator, called *here document.* The operator is followed by any arbitrary string. All following text is regarded as coming from the standard input, until the arbitrary string appears a second time.

Program 4: Printing Text

```
1  #!/bin/bash
2  # save as text.sh
3  # printing text
4  cat <<%%
5  Here comes a lot of text.
6  My home directory is $HOME - cool.
7  Let us use some tabulators:
8          tab1
9          tab2
10 This looks like a nice list.
11
12 %%
13
14 cat <<\%%
15 Here comes a lot of text.
16 My home directory is $HOME - cool.
17 Let us use some tabulators:
18         tab1
19         tab2
20 This looks like a nice list.
21 %%
```

Program 4 illustrates the use of <<. In line 4 %% has been used as *text range indicator*. The cat command gets all the text up to the second occurrence of %% in line 12. From line 14 to 21 we use the same construction. However, the operator << is now preceded by a backslash. Therefore the variable *HOME* will not be expanded but printed as it is: "$HOME".

8.4.3 read and line

Two commands are available to get interactive input into your shell script. The most commonly used command is read. On some very old systems the read command is not available. Then you have to use line instead.
Usually you would first ask the user for the required input. This can easily be done with echo. Then you apply the command read in order to save the user's input in a variable.

Program 5: Change $PWD

```
1  #!/bin/bash
2  # save as chg-pwd.sh
3  # changed the environmental variable PWD
4  echo
5  echo "--------------------------------"
6  echo "PWD is currently set as $PWD"
7  echo
8  echo "Enter a new path and press ENTER"
9  echo "or enter nothing and press ENTER"
10 echo "to leave \$PWD unchanged"
11 read new
12 PWD=${new:-$PWD}
13 echo "PWD is now set to $PWD"
14 echo "--------------------------------"
15 echo -e "\a"
```

Program 5 can be considered user-friendly. It uses some echo commands in order to give information and format the output. In line 11 the program stops and continues only after the (Enter) has been pressed. Thus, before you hit (Enter) you can enter a new path that is to be assigned to PATH. The input is saved in the variable *new*. In line 12 it is checked whether the user provided some input or not. If the user only hits the (Enter) key, then *new* will be empty. In that case, the old value of PWD is saved in PWD. However, if *new* is a non-empty variable, the value of *new* will be assigned to PWD (see Sect. 8.5.1 on page 107). In line 15 the script invokes a system beep (see Sect. 8.4.1 on the page before). Keep in mind that variables are assigned only for the active shell. When you start a script, it runs in its own shell. Thus, after termination of *chg-pwd.sh* PWD is unchanged (check it with echo $PWD). In order to change PWD in the current shell you must start the script with

```
. chg-pwd.sh
```

The dot stands for the shell command `source`. In fact, you could also run the script with `source chg-pwd.sh`. The `source` command prevents the script from running in a new sub shell.

8.4.4 Script Parameters

What make scripts very powerful when compared to *aliases* is the possibility to provide parameters. When a shell script is invoked, it receives a list of parameters (also called arguments) from the command line that invoked it. These parameters can then be used to direct the execution of the script. Let us assume you have DNA sequences saved in many individual files. Now, you want to display all files in your home directory and all subdirectories that contain a certain DNA sequence. The shell command would be

```
find -type f -name "*.dna" -exec grep TATAAT {} \;
```

assuming that you wish to query the sequence *TATAAT* in files ending with *.dna*. It is quite a pain always to enter this line. You could write an `alias` for this command; but do you always query for the same sequence? Certainly not! Furthermore, it would be nice to search in other files, too. For example, it is common to save amino acid sequences in files ending with *.aa*. A good option is to write a shell script! Take a look at Program 6.

Program 6: Find Sequence

```
1 #!/bin/bash -x
2 # save as grep_file_seq.sh
3 # search for a pattern (par2) in *.(par1) files
4 # query home directory and all subdirectories
5 find $HOME -type f -name "*.$1" -exec grep "$2" {} \;
```

Program 6 is quite straightforward. You call the script with

```
grep_file_seq.sh dna TATAAT
```

This means you first enter the script name and then provide two parameters, *dna* and *TATAAT*, respectively. These parameters are internally saved in the variables *$1* and *$2*, respectively. Line 1 of Program 6 indicates that it should be interpreted by the bash shell. We also use the option `-x`. This is a good option for debugging purposes, that is: testing the program. With the option `-x` all executed commands will be displayed (see Sect. 8.8 on page 119). The next 3 lines contain comments. They help us to remember what the script is about. Line 5 contains the command itself. Here, you see the use of the system variable *HOME* and the parameters *$1* and *$2*, respectively. They are used by script as they would be typed in. You can use up to 9 parameters. The content of these parameters is explained in Table 8.2 on the following page.

In Program 2 on page 98 we changed the permission of all files in the home directory and all subdirectories having the extension **.sh*. Now, let us write a

Table 8.2. Variables to access shell parameters

Variable	Value
$1 - $9	The first 9 parameters.
$#	The total number of parameters.
*$**	All the parameters.
$0	The name of the script itself.

script that executes the same task, but only in the currently active directory and only with files we specify. The corresponding script is shown in Program 7.

Program 7: Make Files Executable

```
1 #!/bin/bash
2 # save as chmod_files.sh
3 # adds execution permission for the user
4 chmod u+x $*
```

Program 7 gives you an example of how one can use the variable containing all parameters (*$**) in order to supply a command with all parameters at once.

8.5 Substitutions and Expansions

The main job of the shell is to translate a string of characters given by the user (or as a line of a shell script) into a sequence of tokens for execution. These substitution or expansion processes appear in the following order: brace expansion, tilde expansion, parameter expansion, variable substitution, command substitution, arithmetic substitution and pathname expansion. Let us define a word as a sequence of characters without spaces. Brace expansion is the process of expanding every word contained in a brace expression {*w1,w2,w2*} into a sequence of words where the brace expression is replaced by *w1,w2,w3*, respectively. For example, *abc*{*de,fg,hi*} expands into the sequence of words *abcde abcfg abchi*. Tilde expansion expands every ~ into the path to your home directory. Variable substitution and parameter expansion substitute the value of variables. Variable substitution replaces every expression containing *var* by the value of variable *$var*. Parameter expansion is a kind of conditional substitution. There are many variations of parameter expansion. Command substitution replaces every expression of the form `$(cmd)`, where *cmd* is a command, by the output of the execution of *cmd*. Hence, `$(pwd)` is replaced by the path to the active working directory.
This all sounds probably pretty complicated. Things will become clearer when you see the examples in the following sections.

8.5.1 Variable Substitution

A very useful feature of the shell is variable substitution or variable expansion. Expansion or substitution means that you assign a construction around a variable and this construction will be resolved. We already used variable expansion when we worked with shell script parameters in Section 8.4.4 on page 105 or recalled the value of a variable in Terminal 42 on page 101. Variable expansion is initiated by `${variable}`. Here, *variable* stands for the variable name to be expanded. In the list below, `word` affects in the one or the other way variable expansion.

`${var:-word}`	If the variable *var* is empty, the result of this construction is *word*. Otherwise, the value of *var* is the output. *var* remains unchanged.
`${var:=word}`	As above. However, if the variable *var* is empty, *word* will be assigned to it. Thus, the value of *var* might change.
`${var:+word}`	If *var* is empty, nothing happens and the output is empty. Otherwise, the value of *var* will be substituted with *word*. *var* itself remains unchanged!
`${#var}`	Provides the number of characters of the variable *var*. In case *var* is empty, the output is 0.
`${var:offset:length}`	Expands up to *length* characters of *var*, starting at *offset* (the first character is 0). If *length* is omitted, this construction expands from *offset* to the end of the value of *var*.

There are many more possibilities, which you might want to look up in the manpages. Let us take a look at some examples in the following Terminal.

Terminal 45: Parameter Substitution

```
1  $ enzyme="hydrogenase specific endopeptidase"
2  $ echo $enzyme
3  hydrogenase specific endopeptidase
4  $ echo ${enzyme:=text}
5  hydrogenase specific endopeptidase
6  $ echo ${enzyme:+text}
7  text
8  $ echo $gene
9
10 $ echo ${gene:-gene not discovered}
11 gene not discovered
12 $ echo $gene
13
14 $ echo ${gene:=ATG...TAA}
```

```
15 ATG...TAA
16 $ echo $gene
17 ATG...TAA
18 $ echo ${#gene}
19 9
20 $ echo ${gene:3:3}
21 ...
22 $ echo ${gene:3}
23 ...TAA
24 $
```

The examples given in Terminal 45 should give you an insight into functional aspects of variable substitution. In line 1 we assign the value "hydrogenase specific endopeptidase" to the variable named *enzyme*. In line 4, the value of the variable *enzyme* would be replaced by the text string "text", if *enzyme* were empty, which is not the case. The value of *enzyme* is returned and displayed in line 5. In line 6, the content of the variable *enzyme* is replaced because it is not empty. In line 10 of Terminal 45 the text string "gene not discovered" is returned. However, the empty variable *gene* remains unchanged. In line 14 a value is assigned to the previously empty variable *gene*. In line 18 we check for the size of the content of *gene*, whereas we extract parts of *gene* in lines 20 and 22. In all these cases the value of *gene* itself remains unchanged.

8.5.2 Command Expansion

Very often it is desired to use the output of a command in a shell script. Therefore, you need to call the command and catch its output. Basically, the call of the command is converted into its output – this is called command expansion. The output of commands can be expanded in two different ways. One can either use `$(command)` or graves: `` `command` ``. There is no difference between both methods. The following examples make things a bit clearer.

Terminal 46: Command Expansion

```
1 $ echo $(date)
2 Thu Jul 10 15:29:02 WEDT 2003
3 $ echo `date`
4 Thu Jul 10 15:29:13 WEDT 2003
5 $ files=$(ls)
6 $ echo $files
7 dna.seq enzymes.txt structure.pdb test.txt
8 $
```

In lines 1 and 3 of Terminal 46 the command `date` is executed and its result is returned and printed by `echo`. In line 5 we save the output of the `ls` command in the variable *files*. In the next line we print out the content of *files*.

8.6 Quoting

It is very important to understand how the shell handles quotations, that is the use of double quotes, single quotes and the backslash (escape character). Quoting is used to remove or disable the special meaning of certain characters or words to the shell. Thus, quoting prevents reserved words from being recognized as such, and prevents variable expansion.

8.6.1 Escape Character

The backslash (\) is bash's escape character. It preserves the literal value of the next character that follows, with the exception of the newline character.

Terminal 47: Escaping

```
1 $ echo "Research costs some $s"
2 Research costs some
3 $ echo "Research costs some \$s"
4 Research costs some $s
5 $
```

Terminal 47 shows you an example of the importance of the escape character. In line 1 "$s" is interpreted as the variable *s*. In contrast, in line 3 the shell recognizes the dollar character as a literal.

8.6.2 Single Quotes

Enclosing characters in single quotes (') preserves the literal value of all characters within the quotes. Note: A single quote may not occur between single quotes, even when preceded by a backslash!

Terminal 48: Quoting

```
1 $ echo 'Research costs some $s'
2 Research costs some $s
3 $ echo 'Research costs some \$s'
4 Research costs some \$s
5 $
```

The effect of single quotes is demonstrated in Terminal 48. Note that even the backslash is recognized as literal and not interpreted as escape character.

8.6.3 Double Quotes

Enclosing characters in double quotes (") preserves the literal value of all characters within the quotes, with the exception of the dollar character ($), the grave character (`) and the backslash (\). The dollars character and grave retain their special meaning within double quotes. The backslash retains its

special meaning only when followed by one of the following characters: $, `, ", \, or the newline character. Thus, a double quote may be quoted within double quotes by preceding it with a backslash.
A double-quoted string that is preceded by a dollar character is recognized as a variable name and will be replaced by its value. After replacement, the string is double-quoted.

8.7 Decisions – Flow Control

One of the most important features of programs is the power of flow control. In this section you will learn some fundamental programming constructs. Depending on the "answer" to given "questions", the program decides where to proceed. This is why we talk of *flow control*: you can influence how the program flows through the lines. Every programming language has the capacity of flow control. You can execute loops (repetitively execute commands), analyze cases and so on.

8.7.1 `if...then...elif...else...fi`

If...then is a conditional construct. It allows you to make a test before executing an action (see Fig. 8.1).

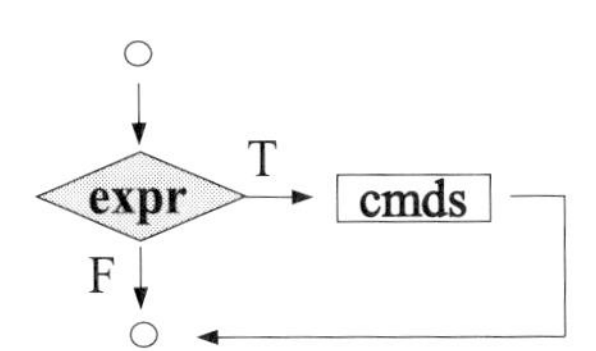

Fig. 8.1. The `if...then` construct. If the expression (*expr*) is true, then execute the command(s) (*cmds*). Typical expressions are discussed in Section 8.7.2 on the next page

The conditional statement is introduced by `if` and evaluates an expression. Depending on the *exit status* of the expression, an action, preceded by `then`, is executed or not. Several such constructs can be grouped (`elif...then`). With `else` an action can be executed if all previous expressions were false. The conditional statement is closed with `fi` (reverse of if).

Thus, the complete syntax is:

```
if expression1; then
  action1
elif expression2; then                              -optional
  action2                                           -optional
else action3                                        -optional
fi
```

The `elif` and `else` commands are optional. The rest is obligatory. If *expression1* returns 0 (i.e., true), then *action1* is executed. If *expression1* does not return 0 (i.e. false), then *expression2* is tested. If *expression 2* returns 0, then *action2* is executed. Else, *action3* is executed. This means that either action1, action2 or action3 is executed. Of course, each action could consist of several actions in separate lines.
Usually commands return the exit status 0 when they have finished their job without any error. This property is used in Program 8.

Program 8: If...Then

```
1  #!/bin/bash
2  # save as if-ls.sh
3  # tests if file is present in current dir
4  # needs 1 parameter
5  if ls $1 >/dev/null 2>&1; then
6    echo "$1 exists"
7  else
8    echo "$1 does not exist"
9  fi
```

Program 8 is executed with

```
./if-ls.sh file
```

The parameter *file* (which is saved in the variable *$1*) is used by the `ls` command in line 5. If the file is present in the active directory, then `ls` will return the exit status 0, else 1. In order not to mess up the output screen with the output of `ls`, we redirect its output and errors (see Sect. 7.5 on page 85) to */dev/null*, the nirvana.
The sequence "*if command1 then command2 fi*" may also be written as "*command1 && command2*" (see Sect. 7.7 on page 88). Conversely, "*command1 || command2*" executes command2 only if command1 returns an exit status other than 0 (false).
A powerful extension for the conditional *if...then* statement is provided by the command `test`, as shown in Section 8.7.2.

8.7.2 `test`

The `test` command, although not part of the shell, is intended for use by shell programs. It can be used for comparisons. For example, "`test -f file`" returns the *exit status* zero if *file* exists and a non-zero exit status otherwise. The exit status of the `test` command (or any other command) can then be analyzed and the behaviour of the program be adapted accordingly. In general, `test` evaluates a predicate and returns the result as its exit status. Some of the more frequently used test arguments are given below. Note that *n1* and *n2* represent different numbers or variables containing numbers, and *s1* and *s2* represent different text strings or variables containing text.

–Numbers–	
`n1 -eq n2`	True if number *n1* is equal to number *n2*.
`n1 -le n2`	True if number *n1* is less than or equal to number *n2*.
`n1 -ge n2`	True if number *n1* is greater than or equal to number *n2*.
`n1 -lt n2`	True if number *n1* is less than number *n2*.
`ni -gt n2`	True if number *n1* is greater than number *n2*.
–Strings–	
`-n s1`	True if string *s1* is not empty.
`-z s1`	True if string *s1* is empty.
`s1 = s2`	True if string *s1* is equal to string *s2*.
`s1 != s2`	True if string *s1* is not equal to string *s2*.
–Files–	
`-z file`	True if *file* exists.
`-f file`	True if *file* is a file.
`-d file`	True if *file* is a directory.
`-r file`	True if *file* is readable.
`-w file`	True if *file* is writable.
`-x file`	True if *file* is executable.
`-s file`	True if *file* is not empty.

With the help of `echo` you can analyze the result of the `test` command in the command line.

Terminal 49: Test

```
1  $ test 2 -eq 2; echo $?
2  0
3  $ test 2 -eq 3; echo $?
4  1
5  $ test $USER; echo $?
6  0
7  $
```

The exit status of `test` is saved in the variable *$?*. Thus, as shown in Terminal 49, you can check the exit status with "`echo $?`". In order to write two commands on one line we have to separate them with a semicolon. If the comparison by `test` returns *true* (exit status 0), then *$?* is zero, and vice versa. The following program gives an example of the combination of *if...then* and test.

```
Program 9: Test Parameters
1  #!/bin/bash
2  # save as test-par.sh
3  # tests if the number of parameters is correct
4  if test $# -eq 1; then
5    echo "Program needs exactly 1 parameter"
6    echo "bye bye"
7    exit 1
8  fi
```

Program 9 checks whether the value of the variable $# is equal to 1. Remember that $# contains the number of parameters (see Sect. 8.4.4 on page 105). Note that the arguments of test are separated by spaces. The if command construct in line 4 is separated from the then command by a semicolon. Otherwise, then must be in a new line. – In modern times, it is very common to invoke test with brackets. Then line 4 would become

```
if [ $# -eq 1 ]; then
```

Note that there are spaces around the brackets!
Another good example is Program 22 on page 122 in Section 8.9 on page 122.

8.7.3 while...do...done

The while construction is used to program loops. A loop is a construct that allows us to execute one or more actions again and again (see Fig. 8.2).

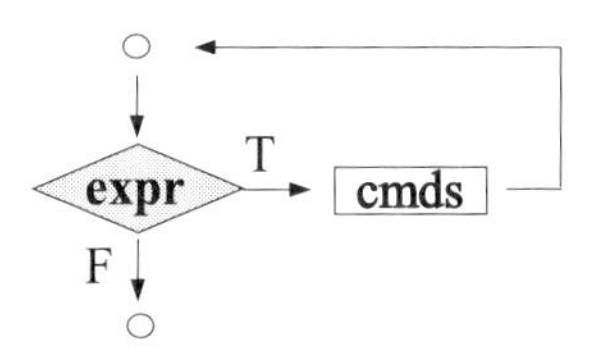

Fig. 8.2. The while...do construct. While the expression (*expr*) returns true, the command(s) (*cmds*) are executed

The loop is aborted when an expression returns the exit status 0 (i.e. the expression becomes false). The loop is introduced with while and ends with done.

The syntax is:

```
while expression; do
  action(s)
done
```

The following program demonstrates the function of *while...do.*

Program 10: While...Done

```
#!/bin/bash
# save as triplet.sh
# splits a sequence into triplets
x=0
while [ -n "${1:$x:3}" ]; do
  seq=$seq${1:$x:3}" "
  x=$(expr $x + 3)
done
echo "$seq"
```

Program 10 expects a DNA sequence from the command line. Thus, the script is executed with

```
./triplet.sh atgctagtcgtagctagctcga
```

The DNA sequence is then split into triplets and printed out. In line 4 we assign the value 0 to the variable x. The important line is line 5. Here we use the brackets to invoke the `test` command (see Sect. 8.7.2 on page 111). We have learned that the option `-n` will return the exit status 0 when a given string is not empty (see Sect. 8.7.2 on page 111). This means that the expression

`[␣-n␣"${1:$x:3}"␣]`

is true, as long as triplets can be copied from the variable *$1* (the command line parameter). Remember that `${a:b:c}` gives c characters from position b of the variable a (see Sect. 8.5.1 on page 107). Thus, line 5 reads: `while $1:$x:3` is not empty, `do` execute the commands up to `done`. In line 6, a triplet and a space character are added to the variable *seq*. Then, the counter variable x is increased by 3 (3 nucleotides = one triplet). Finally, the sequence is displayed.
Together with the command `shift`, the *while* loop can be used to read all command line parameters. With `shift` the command line parameters are shifted. The parameter assigned to *$9* is shifted to *$8*, *$8* to *$7* and so on. The parameter in variable *$1* is lost. An appropriate way to read all command line parameters would be:

Program 11: Read Command Line Parameters

```
#!/bin/bash
# save as para.sh
# prints command line parameters
```

```
4  while [ -n "$1" ]; do
5    echo "\$#=$# - \$0= $0 - \$1=$1"
6    shift
7  done
```

The `while` loop in Program 11 cycles as long as there are parameters in *$1*. The output of the program is the number of remaining parameters from variable *$#*, the name of the script from *$0* and the active parameter from *$1*.

Sometimes you might end up with a loop that does not end: the program hangs. In that case you can quit the program with Ctrl+C. Alternatively, you can program the loop in such a way that it breaks when a certain condition is fulfilled. The corresponding command is `break`. For example, we could stop Program 10 on the facing page when a stop codon is reached.

Program 12: While...Break

```
1  #!/bin/bash
2  # save as triplet-stop.sh
3  # splits a sequence into triplets
4  x=0
5  while [ -n "${1:$x:3}" ]; do
6    seq=$seq${1:$x:3}" "
7    x=$(expr $x + 3)
8    if [ ${1:$x:3} == taa ] || [ ${1:$x:3} == tga ]; then
9      break
10   fi
11 done
12 echo "$seq"
```

Program 12 shows an example of how the execution of a loop can be stopped. The conditional statement *if...then* breaks the *while...do* loop if the current triplet is a *taa* or *tga*. The *or* is represented by the two vertical bars (||). Remember that expressions can also be connected by the logical *and* statement, represented by two ampersands (&&).

8.7.4 until...do...done

Very similar to the *while* loop is the *until* loop. In fact, it is the negation of the former (see Fig. 8.3 on the following page).

An action is not executed `while` a certain condition is given but `until` a certain condition is given. The syntax is:

```
until expression; do
  action(s)
done
```

Program 13 on the next page is the negation of Program 10 on the facing page. Instead of splitting the command line parameter into triplets *while*

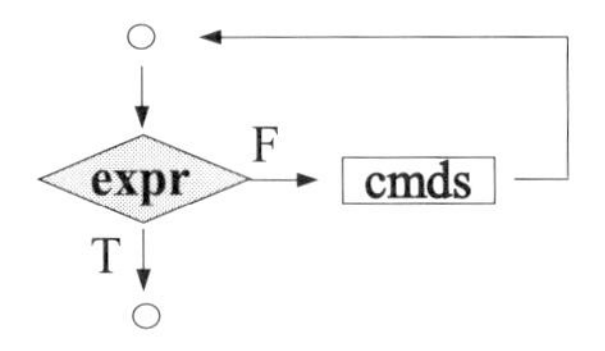

Fig. 8.3. The `until...do` construct. The command(s) (*cmds*) are executed until the expression (*expr*) becomes true

there are still nucleotides, Program 13 splits the sequence until there are no nucleotides left in variable *$1*.

Program 13: Until...Done
```
#!/bin/bash
# save as triplet-until.sh
# splits a sequence into triplets
x=0
until [ -z "${1:$x:3}" ]; do
  seq=$seq${1:$x:3}" "
  x=$(expr $x + 3)
done
echo "$seq"
```

Note the different option for `test`. Here the option is `-z` (returns 0 if the *string is empty*), in Program 10 on page 114 it was `-n` (returns 0 if the *string is not empty*).

8.7.5 `for...in...do...done`

Another useful loop is the *for...do* construct. This is also known as *for* loop. In contrast to the *while* loop, the *for* loop is not dependent on the exit status of an expression but on a list of variables. The syntax is:

```
for variable in list; do              -"in list" is optional
  action(s)
done
```

Let us take a look at a very simple example. I even have the impression that this is the shortest loop script we will write.

Program 14: For Loop
```
#!/bin/bash
# save as for1.sh
# demonstrates for construct
for i in one two three; do
  echo "$i"
done
```

In Program 14 the variable i takes successively the string values "*one*", "*two*" and "*three*". The list could also be the result of a command. For example, all lines of a file could be read with `cat`. In the Program 15 the `ls` command provides a list of all script files.

Program 15: For Loop
```
1 #!/bin/bash
2 # save as for-ls.sh
3 # lists all scripts
4 for i in $(ls *.sh); do
5   echo "File: $i"
6 done
```

If the *for* loop is executed without any list, then the command line parameters are successively called. Thus, Program 11 on page 115 can be run as shown in Program 16.

Program 16: For Loop
```
1 #!/bin/bash
2 # save as for-par.sh
3 # gets command line parameters
4 for i do
5   echo "\$#=$# - \$0=$0 - \$i=$i"
6 done
```

Program 16 does not make use of the `shift` command. The price is that the parameter count *$#* does not decrease its value. In other words, all parameters are kept in their corresponding positional variable *$i*.

8.7.6 `case...in...esac`

As the name implies, you can distinguish different cases with `case`. This is especially helpful when you want to create small user interfaces to guide the user through your program. The case construct is initiated with `case` and ends with `esac` (the reverse of case). The syntax is:

```
case string in
  pattern1)
    action(s)1
  pattern2)
    action(s)2
esac
```

The *string* is compared against the *patterns*. If a pattern matches, the corresponding action will be executed. Again, things get clearer with an example.

Program 17: Case...Esac

```
1  #!/bin/bash
2  # save as case-cp.sh
3  # asks if files shall be backuped
4  # accepts file extension as $1
5  for i in $(ls  *$1); do
6    echo "Backup $i?  yES/nO/qUIT"
7    read answer
8    case $answer in
9      y*) echo "Backup $i"; cp $i $i.bak;;
10     n*) echo "Skip $i";;
11     q*) echo "Quited"; exit;;
12      *) echo "Skip $i";;
13   esac
14 done
```

Program 17 asks you for each file in the current directory whether you want to create a backup or not. The backup is a copy of the file with the extension *.bak*. In order to restrict the file selection, the program accepts a file extension as command line parameter. Thus, if you call the program with

```
./case-cp.sh .sh
```

only script files will be listed. In line 5 the files list is created with a `for` loop. Line 6 asks whether the current file shall be backuped. Your answer is stored in *answer*. Now starts the case discrimination. If *answer* matches $y*$, that is y plus anything, then line 9 is executed. If *answer* matches n plus anything, then line 10 is executed and so on. If *answer* matches neither $y*$, $n*$, nor $q*$, then line 12 is executed.
You can also connect several patterns with a *logical or* (|). Thus, `j*|y*)` would match *yes* and the German *ja*. This can be helpful if you want to create an international interface for your program.

8.7.7 select...in...do

Similar to `case` is `select`. However, with `select` a list is generated from which you can choose. The syntax is as follows:

```
select name in words; do
  action(s)
done
```

The list of *words* following `in` can be either individual words or a variable or a command like `ls` that is expanded, generating a list of items. The list is then displayed on the screen (standard output), each preceded by a number. If `in words` is omitted, the positional parameters supplied with the command line are printed (see Sect. 8.4.4 on page 105). During execution a prompt appears and you are requested to enter a number and press (Enter). You can display a text like "Enter number: " by assigning the message to the variable *PS3*. If the entered number corresponds to one preceding the displayed words, then the value of the variable *name* is set to that word. If the line is empty, the words and prompt are displayed again. Any other value causes name to be set to null. The input is saved in the variable *REPLY*. After each selection, the list of words is displayed again and again, until a `break` command is executed. Let us take a look at one example.

Program 18: Select

```
1 #!/bin/bash
2 # save as select.sh
3 # demonstrates select
4 PS3="Select item: "
5 select name in Protein DNA RNA ; do
6   echo "You selected $name"
7   break
8 done
9 echo "bye bye"
```

In line 4 of Program 18 we assign the value "Select item: " to the variable *PS3*. With this setting the user will be asked to select a item when the list is displayed. Here, the list consist of the three words "Protein DNA RNA". When you run the script it will look as in Terminal 50.

Terminal 50: Select

```
1 $ ./select.sh
2 1) Protein
3 2) DNA
4 3) RNA
5 Select item: 3
6 You selected RNA
7 bye bye
8 $
```

With line 7 of Program 18 you force the *select...done* loop to break and continue the script after the `done` statement. In our case we just print "bye bye".
Like the `case` construct, the `select` construct is well suited to build a simple user interface for your scripts. Such user interfaces are often highly welcome by the user because they facilitate intuitive usage of your script.

8.8 Debugging

A very import part of programming is debugging. If there is an error in your program, this is called a *bug*. The process of searching and removing any bug is called *debugging*. Of course, there is no automatic procedure to remove bugs. You have to go through your program line by line and identify what is wrong. This is the really time-intensive part of programming.

8.8.1 bash -xv

The shell offers you some support for debugging. When you start the shell with the option `-x`, it will print each command it is executing. Thereby, you might identify the malleus line. With the option `-v`, each line of the script is printed before execution. Thus, you see the original command without any command expansion [like `$(command)`] or parameter substitution.

Program 19: Debugging

```
1 #!/bin/bash -vx
2 # save as date-vx.sh
3 # demonstrates debugging
4 date
5 lss
6 now=$(date)
7 echo "bye bye"
```

In the first line of Program 19 we evoke the bash shell with the options `-vx`. The commands of the script should be clear. In line 4 we call the command `date`, in line 5 the non-existing command `lss`, in 6 we assign the current date to the variable *now* and in line 7 we print *bye bye*. Now let us take a look at what we see when we execute the program.

Terminal 51: Debugging

```
1  $ date-vx.sh
2  #!/bin/bash -vx
3  # save as date-vx.sh
4  # demonstrates debugging
5  date
6  + date
7  Son Mai 25 11:16:44 CEST 2003
8  lss
9  + lss
10 ~/scripts/date-x.sh: line 5 lss: command not found
11 now=$(date)
12 date
13 ++ date
14 + now=Son Mai 25 11:16:44 CEST 2003
15 echo "bye bye"
16 + echo 'bye bye'
17 bye bye
18 $
```

Terminal 51 shows the output of Program 19 on the preceding page. You see that all program lines are printed onto the screen. Commands that are executed are preceded by either + or ++. You can also see the error message from the non-existent `lss` command.
When you write larger scripts, you might get confused by all the output. Thus, you might want to restrict the error search to a certain part of your program. This is possible with the `set` command.

Program 20: Debugging

```
1  #!/bin/bash
2  # save as date-set.sh
3  # demonstrates debugging
4  date
5  lss
6  set -vx
7  now=$(date)
8  set +vx
9  echo "bye bye"
```

In line 6 of Program 20 we activate the options `-vx`, in line 8 we deactivate them. By doing this, we restrict debugging to line 7.

8.8.2 `trap`

Some signals cause shell scripts to terminate. The most common one is the interrupt signal Ctrl+C typed while a script is running. Sometimes a shell script will need to do some cleanup, such as deleting temporary files, before exiting. The `trap` command can be used either to ignore signals or to catch them to perform special processing. For example, to delete all files called **.tmp* before quitting after an interrupt signal was received, use the command line

```
trap 'rm *.tmp; exit' 2
```

The interrupt signal Ctrl+C corresponds to signal 2. If this signal is received, two commands will be executed: `rm *.tmp` and `exit`. You can make a shell script continue to run after logout by letting it ignore the hang up signal (signal 1). The command

```
trap ' ' 1
```

allows shell procedures to continue after a hang up (logout) signal.

Program 21: Trap Signal

```
1  #!/bin/bash
2  # save as trap.sh
3  # catch exit signal
4  while true; do
5   echo "test"
6   trap 'echo "bye bye"; exit' 2
7  done
```

Be careful when you enter Program 21. `while` always gets the result `true`. Thus, we initiate an endless loop. The screen will fill up with the message *test*. However, when you press Ctrl+C, line 6 is executed. It *traps* the termination signal, prints `bye bye` and exits the script. You should be very careful when you use `trap`. It is always a good option to terminate a program. However, if you misspell, for example, `exit`, then your program will hang.

8.9 Examples

There is no other way to learn than by doing. In order to practise the function of shell scripts you will find a number of examples in this section. Save these scripts and execute them. Then start to change things and see how the scripts behave.

8.9.1 Check for DNA as File Content

Program 22 tests whether a file contains the characters *actg* or not.

Program 22: Check File Content

```
1  #!/bin/bash
2  # save as dna-test.sh
3  # test if file contains dna sequence
4  # file in $1
5  if test -z $(cat $1) && test  -f $1; then
6    echo "File $1 is empty"
7    exit
8  fi
9  grep -sq '[^acgt]' $1
10 result=$?
11 if test $result -eq 0 ; then
12   echo "File $1 does not contain pure DNA sequence"
13 elif test $result -eq 1; then
14   echo "File $1 does contain pure DNA sequence"
15 elif [ $result -eq 2 ]; then
16   echo "File $1 does not exist"
17 else
18   echo "Some error occured!"
19 fi
```

You call the script with the name of the file that you want to analyze:

```
./dna-test-sh seq.dna
```

In line 5 of Program 22 there are two tests connected by `&&` (logical *and*). This means that lines 5 and 6 are executed only if both tests return *true*. Alternatively, one might want to use the logical *or*, which is "||". The first test in line 5 checks if the file content is empty; the second test checks if the file exists. If one of these two conditions is false, the message in line 6 will be displayed and the program exits. Otherwise, the `grep` command in line 9 checks if the file given with the parameter *$1* contains characters other than *actg*. Depending on the result, `grep` finishes its job with different exit states. If `grep` has found other characters than *acgt*, then the exit status saved in *$?* is 0, or else it is 1. If the file does not exist, then the exit status of `grep` is 2. The exit status is saved in the variable *result* in line 10. Then the value of *result* is tested in a row of *if...then...elif...else...fi* as described in Section 8.7.1 on page 110. In line 15 an alternative way to call `test` is shown: the command can be replaced by using brackets `[ ]`.

8.9.2 Time Signal

The following program "beeps" the number of hours. When you run the script at 4 o'clock, you will hear 4 system beeps. You can call the script from the cron daemon every hour in order to have a time signal.

Program 23: Time Signal

```
1  #!/bin/bash
2  # save as Time-Signal.sh
3  # gives a time signal every hour when connected to cron
4  time=$(date +%I)
5  count=0
6  while test $count -lt $time; do
7    echo -e "\a"
8    sleep 1 # sleep for one second
9    count=$[$count+1]
10 done
```

The heart of Program 23 lies in line 4. "`$(date +%I)`" returns the current hour in 12-hour format (01–12). This value is saved in the variable *time* and used in the `while` loop spanning from lines 6 to 10. The beep signal is generated by the escape sequence "`\a`" in line 7 (see Sect. 8.4.1 on page 103).

8.9.3 Select Files to Archive

This script asks the user for each file in the active directory whether it should be added to an archive or not. Finally, all selected files are archived in a file called *archive*.

Program 24: Interactively Archive Files

```
1  #!/bin/bash
2  # save as archive-pwd-i.sh
3  # interactively archive files with tar
4  array=($(ls))
5  count=0
```

```
6   while test $count -lt ${#array[*]}; do
7     echo "Archive ${array[count]}?"
8     echo " press Enter = no"
9     echo " press y & Enter = yes"
10    read input
11    case $input in
12      y*) list="${list} ${array[count]}"
13    esac
14    count=$(expr $count + 1)
15  done
16  echo "Files: $list"
17  echo "have been added to the file archive"
18  tar -cf archive $list
```

Program 24 demonstrates the use of command expansion. In line 4 the output of the command `ls` is resolved and saved into an array. Here, an array is a 2-dimensional variable, like a staple of paper sheets. Each sheet contains some information. Variable 1 is *array[0]* that contains the first file delivered by `ls`. Variable 2 is *array[1]* that contains the second file of the directory as value and so on. With $#array[*] we can check how many elements the array has. In line 6 we use this statement for a loop. The loop `while` to `done` is repeated until the command `test` gives the output "false", that is, until the variable `count` is less than (`-lt`) the size of the array determined by `${#array[*]}`. What is going on in the *while...done* loop? In line 7 the user is asked whether the file that is saved in the array element `array[count]` should be archived. The user must answer the question as stated in lines 8 and 9. Then follows the `read` command (see Sect. 8.4.3 on page 104). At this point the program stops and waits for user input. The program resumes execution when the user presses (Enter). The text typed by the user is saved in the variable *input*. Between lines 11 and 13 a case discrimination is performed. It starts with `case` and ends with `esac` (the reverse of *case*). In line 11 the program checks whether the variable *input* is equal to something in the following lines, up to the `esac` command in line 13. The case we are interested in is whether the user typed *y* or *yes*. This can be expressed by `y*`: any word that starts with *y*. If *input* matches that case, then the command following `y*)` is executed: here we add the file to the variable *list*. The command reads like: save the content of the variable *list*, plus a space character (␣), plus the current content of the array element `array[count]` in the variable *list*. In line 14 we increase the content of the variable *count* by one. In order to perform an arithmetical calculation we use the command `expr` (expression). That tells the shell to calculate the rest of the line instead of just adding the character string "+ 1" to the variable *count*. When the value of the variable *count* exceeds the number of files in the current directory (saved in *array[ ]*), then the *while* loop stops and the program continues with line 16. Here the list of files to be archived is printed out. The archiving command `tar` itself is executed in line

18. All files chosen by the user to be archived is provided with the variable *list*.

8.9.4 Remove Spaces

The following program converts spaces in filenames into underscores. Especially filenames created on computers running Windows tend to contain spaces. In principle, both Unix and Linux have no problems with spaces in file names; however, you have to escape them with the backslash character. This is rather uncomfortable. If you have a number of files that need to be converted, then you will appreciate the following script.

Program 25: Convert Spaces in Filenames to Underscores

```
1  #!/bin/sh
2  # save as space-convert.sh
3  chgname() {
4     echo "$1" | sed -e 's/[ ][ ]*/ /g
5        s/[ ]/_/g'
6  }
7  find . -name '* *' | sort | while read name do
8     file=`basename "$name"`
9     stem=`dirname "$name"`
10    nfile=`chgname "$file"`
11    nstem=`chgname "$stem"`
12    if [ "$file" != "$nfile" ]
13    then
14       mv "$stem/$file" $nstem/$nfile
15    fi
16 done
```

Program 25 begins with defining a user function in line 3. It pipes filenames saved in *$1* to a little `sed` script. This script actually transforms space characters into one single underscore character. You will learn more about `sed` in Chapter 10 on page 141. The `find` command in line 7 queries the active directory for files containing space characters. These are sorted by `sort` and finally saved in the variable *name*. The shell command `basename` extracts the filename of *name*, whereas `dirname` extracts the path of the file saved in *name*. In lines 10 and 11 the user function `chgname` is called. In line 14 the original files are renamed.

Exercises

The following exercise sounds easier than it is...

8.1. Go through all the programs and examples in this section and play around. Modify the code and observe the changes.

9

Regular Expressions

I am sure you have more than once used the internet search engine *Google*. You have not? Then you probably used another search engine – there are plenty of them out there in the web space. As with all searches, the problem is *finding*. The better you define your search problem, the better your query result will be. In this chapter you will learn how to query for text patterns. Some definitions before we start: we are going to use the terms *literal*, *metacharacter*, *target string*, *escape sequence* and *search pattern*. Here comes a definition of these terms:

Literal	A literal is any actual character that we use in our search, for example, to find *inu* in *Linux*. *inu* is a literal string – each character plays a part in the search; it is literally the string we want to find.
Meta Character	Meta characters are one or more special characters that have a unique meaning and are not used as literals in the search. For example, the dot character "." is a meta character.
Escape Sequence	An escape sequence is a way of indicating that we want to use one of the meta characters as a literal. In a regular expression an escape sequence involves placing a backslash (\) in front of the meta character. In order to find the dot character we have to use the escape sequence "\.".
Target String	The target string defines the sequences of characters we are searching for. In other words, the target string is the string we want to match in the source file.
Search Pattern	The search pattern, or construct, describes the expression that we are using in order to search our target string.

A regular expression is a set of characters that specify a *search pattern*. They are used when you want to search for lines of text containing a particular sequences of characters, the *target string*. It is simple to search for a specific word or string of characters. Almost every text editor, like Microsoft Word, on any computer system can do this. Unlike simple text queries, regular expressions allow you to search for text which matches a particular pattern. You can search for words of a certain size, or for a word with three *t*s. Numbers, punctuation characters, DNA sequences, you name it, a regular expression can find it. Regular expressions are always used in conjunction with a program. What happens once the program you are using has found the pattern is another matter. Some just search for the pattern and print out the line containing the pattern (`egrep`). Editors can replace the string with a new pattern (`vi`). It all depends on the utility. Regular expressions look a lot like the file-matching patterns (wildcards) the shell uses (see Sect. 7.10 on page 91). Sometimes they even act in the same way: the brackets are similar, and the asterisk (*) acts similarly, too. However, regular expressions are much more powerful than wildcards.
Remember that filename wildcards are expanded before the shell passes the arguments to the program. To prevent this expansion, *regular expression must be embedded within single quotes.* The following examples give you a first impression of how you can find target strings with regular expressions.

Search Pattern	*Target String*
`(DNA|RNA)`	Search for "DNA" or "RNA".
`(D|R)NA`	Search for "DNA" or "RNA".
`[tmr]RNA`	Search for "tRNA", "mRNA", or "rRNA".
`*omics`	Search for "Genomics", or "Protenomics", or any other "*nomics*".
`b.g`	Matches any character between "b" and "g", like big, bug, bag...
`[a-zA-Z]`	Matches any single lowercase or uppercase character.
`[^a-zA-Z0-9]`	Matches any single character which is not a number or letter.
`^[eE]nzyme`	Matches the words "Enzyme" or "enzyme" at the beginning of a line.
`[eE]nzyme$`	Matches the words "Enzyme" or "enzyme" at the end of a line.

As you will learn soon, there are much more powerful queries possible.

9.1 Using Regular Expressions

Regular expressions do not work on their own. They are used together with a command or programming language such as `grep`, `find`, `vi`, `sed`, `awk`, `perl`,

MySQL (a database system), *Javascript*, *Java*, *PHP* and many, many more. If you want to use `grep` with regular expressions, you must run it with the option `-e`. Most Unix/Linux systems have a built-in alias, which is the command `egrep`. This means, using `egrep` is the same as using `grep -e`.
There is nothing easier than having the wrong regular expression to find a target string. It really needs some exercise in order to figure out how to make it right. Before you apply a search pattern for an important task, you should check it with a small self-made test file. It is important to check if the regular expression really finds the desired target and that it excludes all non-wanted targets. This becomes especially important when we start to use regular expressions for text replacements. Test your regular expression thoroughly. As you will see, `egrep` (used in this section) and `vi` (see Sect. 9.3 on page 139) are good tools for doing this.

As described above, regular expressions consist of two character types: *meta characters* and *literals*. Meta characters are special characters like wildcards. All other characters are literals. In a way, meta characters describe a desired arrangement of the literals, that is the target string. For the following examples we will use the file *structure.pdb*. It is a cut-down version of a file from the *Brookhaven Protein Data Bank* entry *1CFZ* describing the crystal structure of a protein. The complete file can be downloaded from *www.rcsb.org/pdb*.

```
structure.pdb
1  HEADER  Hydrogenase                              23-Mar-99    1CFZ
2  COMPND  Hydrogenase Maturating Endopeptidase Hybd From
3  SOURCE  ORGANISM_SCIENTIFIC: Escherichia coli
4  AUTHOR  Fritsche, Paschos, Beisel, Boeck & Huber
5  REMARK NCBI PDB FORMAT VERSION 5.0
6  SEQRES 1 A  162  MET ARG ILE LEU VAL LEU GLY VAL GLY ASN
7  SEQRES 2 A  162  THR ASP GLU ALA ILE GLY VAL ARG ILE VAL
8  SEQRES 3 A  162  GLU GLN ARG TYR ILE LEU PRO ASP TYR VAL
9  SEQRES 4 A  162  ASP GLY GLY THR ALA GLY MET GLU LEU LEU
10 HELIX  1 hel ILE A   18  GLN A   28
11 HELIX  2 hel PRO A   92  THR A  107
12 HELIX  3 hel ILE A  138  SER A  152
13 SHEET  1 str ARG A    2  ASN A   10
14 SHEET  2 str ARG A   29  LEU A   32
15 SHEET  3 str TYR A   35  THR A   43
16 ATOM 1  C    MET A 1 48.865  25.394  51.393  1.00 54.58 C
17 ATOM 2  CA   MET A 1 49.879  24.359  50.932  1.00 59.61 C
18 ATOM 3  CB   MET A 1 49.248  23.457  49.877  1.00 62.37 C
19 ATOM 4  CE   MET A 1 51.349  24.403  47.765  1.00 71.39 C
20 ATOM 5  CG   MET A 1 48.708  24.106  48.629  1.00 66.70 C
21 ATOM 6  N    MET A 1 50.347  23.578  52.116  1.00 62.03 N
22 ATOM 7  O    MET A 1 47.875  25.011  52.020  1.00 54.99 O
23 ATOM 8  SD   MET A 1 49.731  23.948  47.163  1.00 77.15 S
```

You might ask: what is the content of this file? A protein can be characterized by three important features: its amino acid sequence (i.e. primary structure), the presence of α-helices and β-sheets as structural building blocks (secondary structure) and the overall three-dimensional structure (tertiary structure). With methods like X-ray or NMR spectroscopy one can resolve the tertiary structure of a protein. This means that to each atom, a position in space can be assigned. You know that it is the three-dimensional structure that determines the function of a protein.
All structural information of a protein is saved in a special file format: the *Brookhaven Protein Data Bank File Format.* Our example file resembles this file format. However, the length of fields has been cut down for printing purposes. The first 5 lines of *structure.pdb* give some background information, the next 4 lines contain the protein sequence, followed by secondary structure features (lines 10–15) and the position of the atoms. Generally, the file content is organized in lines. The content of each line is indicated by the first word (HEADER, COMPND, SOURCE, AUTHOR,...). Again, please note that both lines and rows of the original file have been truncated!

9.2 Search Pattern and Examples

In this section you will learn more about different types of regular expressions and how to apply them. Here, we will work with the `grep` command that understands regular expressions, that is `egrep`. `egrep` works in the same way as `grep` does (see Sect. 6.1.5 on page 68): it searches the input (usually a file) for a specified text pattern. The input text is treated in lines. Lines that contain the query pattern are printed to the standard output (the screen). In contrast to `grep`, the query pattern used with `egrep` may contain regular expressions. Remember: The search pattern must be enclosed in single quotes ('...'). This prevents the shell from executing substitutions. Our first example file will be *structure.pdb* from Section 9.1 on page 128.

9.2.1 Single-Character Meta Characters

A typical way to write the name of an amino acid is the *three-letter code.* The amino acids glycine and glutamine are represented by GLY and GLN, respectively. Thus, they differ by only one character. There are different ways to query for such differences. The meta characters used in such queries are called *single-character meta characters.* These include the following symbols:

`.`	Matches any single character except the newline character.
`[ ]`	Matches any character listed between the brackets. `[acgt]` matches "a" or "c" or "g" or "t". The four characters ".", "*", "[" and "\" stand for themselves within such a list of characters.
`[^ ]`	Matches any character except those listed between the brackets.
`[0-9]`	Matches any number in the *range* between *0* and *9*. The dash (`-`) indicates a range of of consecutive characters. `[0-3a-cz]` equals `[0123abcz]`.

Okay, let us see how we can apply these meta characters in regular expressions. Let us assume that we want to match all lines containing *GLU* and *GLN* in *structure.pdb*. The appropriate command is shown in Terminal 52.

Terminal 52: egrep GL.

```
1  $ egrep 'GL.' structure.pdb
2  SEQRES 1 A  162  MET ARG ILE LEU VAL LEU GLY VAL GLY ASN
3  SEQRES 2 A  162  THR ASP GLU ALA ILE GLY VAL ARG ILE VAL
4  SEQRES 3 A  162  GLU GLN ARG TYR ILE LEU PRO ASP TYR VAL
5  SEQRES 4 A  162  ASP GLY GLY THR ALA GLY MET GLU LEU LEU
6  HELIX  1 hel ILE A   18  GLN A   28
7  $ egrep 'GLN' structure.pdb
8  SEQRES 3 A  162  GLU GLN ARG TYR ILE LEU PRO ASP TYR VAL
9  HELIX  1 hel ILE A   18  GLN A   28
10 $ egrep 'GLY' structure.pdb
11 SEQRES 1 A  162  MET ARG ILE LEU VAL LEU GLY VAL GLY ASN
12 SEQRES 2 A  162  THR ASP GLU ALA ILE GLY VAL ARG ILE VAL
13 SEQRES 4 A  162  ASP GLY GLY THR ALA GLY MET GLU LEU LEU
14 $
```

Remember that the regular expression "`GL.`" in line 1 of Terminal 52 must be embedded in single quotes. "`GL.`" will match every line of the file *structure.pdb* that contains anywhere the succession of the characters *G* and *L* followed by any other character, which is represented by the dot (.). The matching lines are printed onto the screen. Lines 7 and 10 show the result of the search for the individual amino acids *GLN* and GLY.
Now let us search for the amino acids threonine (THR) and tyrosine (TYR). The correct command would be:

```
egrep 'T.R' structure.pdb
```

All lines where an amino acid lies between alanine (ALA) and glycine (GLY) are matched with

```
egrep 'ALA.....GLY' structure.pdb
```

In this example the five dots represent any five consecutive characters, including the spaces. The command matches exactly one line, which is line 7. Alternatively, you could match the same pattern with

```
egrep 'ALA␣...␣GLY' structure.pdb
```

This example shows that the spaces are treated as normal characters.
The brackets represent a single character from a selection of characters. This selection is contained within the brackets. In our file three amino acids start with *GL*: glycine (GLY), glutamine (GLN) and glutamic acid (GLU). How can we match these lines containing either GLU or GLY? Now we employ a selection.

Terminal 53: Brackets

```
1  $ egrep 'GL[YU]' structure.pdb
2  SEQRES 1 A  162  MET ARG ILE LEU VAL LEU GLY VAL GLY ASN
3  SEQRES 2 A  162  THR ASP GLU ALA ILE GLY VAL ARG ILE VAL
4  SEQRES 3 A  162  GLU GLN ARG TYR ILE LEU PRO ASP TYR VAL
5  SEQRES 4 A  162  ASP GLY GLY THR ALA GLY MET GLU LEU LEU
6  $
```

The structure `[YU]` in Terminal 53 represents one character: either *Y* or *U*. The list could be much longer. The list may also contain a range of characters indicated by a dash (-).

Terminal 54: Character List

```
1  $ egrep '[0-2T] A' structure.pdb
2  SEQRES 1 A  162  MET ARG ILE LEU VAL LEU GLY VAL GLY ASN
3  SEQRES 2 A  162  THR ASP GLU ALA ILE GLY VAL ARG ILE VAL
4  ATOM 1  C   MET A 1 48.865  25.394  51.393  1.00 54.58 C
5  ATOM 2  CA  MET A 1 49.879  24.359  50.932  1.00 59.61 C
6  ATOM 3  CB  MET A 1 49.248  23.457  49.877  1.00 62.37 C
7  ATOM 4  CE  MET A 1 51.349  24.403  47.765  1.00 71.39 C
8  ATOM 5  CG  MET A 1 48.708  24.106  48.629  1.00 66.70 C
9  ATOM 6  N   MET A 1 50.347  23.578  52.116  1.00 62.03 N
10 ATOM 7  O   MET A 1 47.875  25.011  52.020  1.00 54.99 O
11 ATOM 8  SD  MET A 1 49.731  23.948  47.163  1.00 77.15 S
12 $
```

In Terminal 54 the pattern "`[0-2T]␣A`" matches one of the characters *0*, *1*, *2* or *T*, followed by the space character (␣), followed by the character *A*. In order to invert a selection, it is preceded by a circumflex (^). Thus,

```
egrep '[^0-2T]␣A' structure.pdb
```

would match every line where the pattern "`[0-2T]␣A`" does *not* match.
Up to now we have used only a single-character pattern. Now let us see how we can match repetitions of a particular pattern.

9.2.2 Quantifiers

The regular expression syntax also provides meta characters which specify the number of times a particular character should match. Quantifiers do not work on their own but are combined with the single character-matching patterns discussed above. Without quantifiers, regular expressions would be rather useless. The following list gives you an overview of quantifiers:

`?`	Matches the preceding character or list zero or one times.
`*`	Matches the preceding character or list zero or more times.
`+`	Matches the preceding character or list one or more times.
`{num}`	Matches the preceding character or list *num* times.
`{num,}`	Matches the preceding character or list at least *num* times.
`{min,max}`	Matches the preceding character or list at least *min* times, but not more than *max* times. The numbers must be less than 65536 and the first must be less than or equal to the second.

It is quite easy to get confused with these expressions, especially with the first three. The most universal quantifier is the star (`*`). In combination with the dot (`.`) it matches either no character or the whole line. As mentioned above, quantifiers are not used on their own. They always refer to the preceding character or meta character. Let us look at some examples with a new test file named *sequence.dna*. This file contains two arbitrary DNA sequences in the *FASTA format*. Sequences in FASTA format have a sequence name preceded by the ">" character. The following line(s) contain the sequence itself.

sequence.dna
```
1 >seq1 the first test sequence
2 ATGxxxTAAxxATGxxTAAGACGCTAGCTCAGCATCGACTACGATCCT
3 GATAGCTATGTCGATGCTGATGCATGCATGCGGGGGGATTGAAAAAGG
4 CGTGTGTAGCGTAATATATGCTATAGCATTGGCATTA
5
6 >seq2 the 2nd test sequence
7 AGCGGCGGCAGTACTGCTATTCGATGTACGGCGATATGCATGGGGATT
8 TAATAAACACAATGCGGTGTAGGGGGAAAAATTTnAGCATGCAATT
```

A little repetition: how do we get all lines containing sequence names filtered out? Just use

```
egrep '>' sequence.dna
```

Now let us match lines having a start codon (*ATG*) followed by some nucleotides and a stop codon (*TAA*):

```
egrep 'ATG.*TAA' sequence.dna
```

The next regular expression shows all lines containing two or three, but not more, consecutive *A*s:

```
egrep '[CGT]AAA?[CGT]' sequence.dna
```

The questions mark indicates that the preceding *A* might be present once or not present at all in the matching pattern. The `[CGT]` at the beginning and end of the regular expression prevents that a row of more than three *A*s is recognized. This could also have been achieved with `[^A]`, representing any character but *A*:

```
egrep '[^A]AAA?[^A]' sequence.dna
```

If you want to detect all lines with two or more repeated *A*s use

```
egrep '[^A]AA+[^A]' sequence.dna
```

The plus character indicates that the preceding *A* should be present at least one time. By using braces ({ }) you can exactly define the number of desired repetitions. Thus, to find a repetition of 4 *A*s or more, use

```
egrep 'A{4,}' sequence.dna
```

If you want to search for braces you have to escape them with a backslash (see Sect. 9.2.5 on the next page)!
As you can see from these examples, regular expressions are really powerful pattern-matching tools.

9.2.3 Grouping

Often it is very helpful to group a certain pattern. Then you have to enclose it in parentheses: (...). In this way you can easily detect repeats of the sequence *AT*:

```
egrep 'AT(AT)+' sequence.dna
```

Note that if we left away the first *AT*, we would also match single occurrences of *AT*, though we are looking for repeats, that is more than one occurrence. The next example searches in the structure file *structure.pdb* for repeating glycines:

```
egrep '(GLY␣){2,}' structure.pdb
```

Note that there is a space character after *GLY*. This is necessary because the amino acids are separated by spaces! In order to query for parentheses you must escape them with a preceding backslash (see Sect. 9.2.5 on the facing page).

9.2.4 Anchors

Often you need to specify the position at which a particular pattern occurs. This is often referred to as *anchoring* the pattern:

^	Matches the start of a line.
$	Matches the end of a line.
\<	Matches the beginning of a word.
\>	Matches the end of a word.
\b	Matches the beginning or the end of a word.
\B	Matches any character not at the beginning or end of a word.

Let us come back to the file *structure.pdb*. Find each line that begins with the character *H*. The correct command is

```
egrep '^H' structure.pdb
```

Now, you can also easily search for empty lines:

```
egrep '^$' sequence.dna
```

If you want to match all empty lines plus all lines only containing spaces, you use

```
egrep '^␣*$' filename
```

With this command you see only empty lines on the screen. More probable is the situation that you want to print all non-empty lines. That can be achieved with

```
egrep '[^␣]' filename
```

Empty lines, or lines containing only space characters, do not match this regular expression. Again, note that the space character is treated as any other literal. In regular expressions the space character cannot be used to separate entries!
Regular expressions will also help us to format the output of the `ls` command. How can we list all directories? This task can easily be achieved by piping the output of "`ls -l`" to `egrep`:

```
ls -l | egrep '^d'
```

`egrep` checks whether the first character of the file attributes is a "d" (see Sect. 4.2 on page 38). In a similar manner you could list all files that are readable and writable by all users:

```
ls -l | egrep '^.{7}rw'
```

Here the search pattern requires that the 8th and 9th character of a line equals "r" and "w", respectively.

9.2.5 Escape Sequences

By now, you are probably wondering how you can search for one of the special characters (asterisks, periods, slashes and so on). As for the shell, the answer lies in the use of the escape character, that is the backslash (\). In order to override the meaning of a special character (in other words, to treat it as a literal instead of a meta character), we simply put a backslash before that character. Thus, a backslash followed by any special character is a one-character regular expression that matches the special character itself. This combination is called an *escape sequence*. The special characters are:

`. * [ \`	Period, asterisk, left square bracket and backslash, respectively, which are always special, except when they appear within brackets ([ ]).
`^`	Caret or circumflex, which is special at the beginning of an entire regular expression, or when it immediately follows the left bracket of a pair of brackets ([ ]).
`$`	Dollar character, which is special at the end of an entire regular expression.

Wrong application of escape sequences, that is wrong escaping, is the number one error when using regular expressions. Be aware of this fact and always test your construct before you are starting to do important things with it. Note: Depending on the program with which you are using regular expressions, meta characters like braces and parentheses must be escaped. This is, for example, the case in `sed` (see Sect. 10.4.2 on page 147).

9.2.6 Alternation

In the first example in Terminal 52 on page 131 we wanted to match both GLY and GLN. You will now learn an alternative way to achieve this: alternation. Alternation refers to the use of the "|" character to indicate logical OR. Commonly, the alternation is used with parentheses to limit the scope of the alternative matches. Consider the following regular expression, which accounts for both GLY and GLN:

```
egrep '(GLY|GLN)' structure.pdb
```

This construction will give you the same result as is shown in Terminal 52 on page 131. You can also combine more of these statements as shown in the following example:

```
egrep '(GLY|GLN|ILE)' structure.pdb
```

There is one point to keep in mind when you combine even more expressions. The regular expression "`one and|or two`" is equal to "`(one and)|(or two)`" but not equal to "`one (and|or) two`"! You are usually on the safe side by using parentheses!

9.2.7 Back References

In Sections 9.2.3 on page 134 and 9.2.6 on the facing page you have already seen the use of parentheses in order to group constructs. However, the constructs within parentheses are not only grouped but also memorized in an internal memory. This means that you can refer to previously found patterns within your construct. The following meta characters take care of back referencing.

`( )` Memorizes the match for the regular expression enclosed in parentheses.

`\n` Recalls the *n*th match. The match for the first construct enclosed in parentheses is recalled by `\1`, the match for the second construct enclosed in parentheses is recalled with `\2` and so forth.

The application of back referencing becomes clearer with the following example. Assume we wish to find every line containing the repetition of a character. Terminal 55 shows you the solution.

Terminal 55: Memory

```
1  $ egrep '([A-Za-z])\1' structure.pdb
2  SHEET  1 str ARG A    2  ASN A  10
3  SHEET  2 str ARG A   29  LEU A  32
4  SHEET  3 str TYR A   35  THR A  43
5  $
```

The construct "`[A-Za-z]`" enclosed in parentheses in line 1 of Terminal 55 matches any alphabetic character, no matter whether it is upper- or lowercase. Then we recall that match with "`\1`". Thus, the whole construct matches every repetition of any characters.
With the next example we will find DNA sequence repeats in the file *sequence.dna*.

Terminal 56: DNA Repeats

```
1  $ egrep '([ACTG][ACTG])\1\1' sequence.dna
2  GATAGCTATGTCGATGCTGATGCATGCATGCGGGGGATTGAAAAAGG
3  CGTGTGTAGCGTAATATATGCTATAGCATTGGCATTA
4  $
```

The regular expression in line 1 of Terminal 56 matches repeats of DNA nucleotide duplets. The match in line 2 is *GGGGGG* whereas the match in line 3 is *GTGTGT* and *ATATAT*. With the following construct

```
egrep '([ACTG])([ACTG])([ACTG])\3\2\1' filename
```

we want to match inverted repeats of triplets, such as *TCAACT*, *GCGGCG* or *AAAAAA*. Note that you cannot use the construct

```
egrep '([ACTG]){3}\3\2\1' filename
```

for the same purpose. You would see the error message "bad back reference". The reason is that the parentheses must be really present. You cannot *virtually* repeat them with a quantifier.
Unfortunately, it is not possible to indicate where the pattern occurs in the matching line. This a limitation of `egrep` and we will overcome this limitation later with `sed`. Another limitation is that we cannot detect patterns stretching over two or more lines. `egrep` works through the file line by line and only analyses single lines. Later we will see examples of how to overcome this limitation, too.

9.2.8 Character Classes

As we saw in Section 9.2.1 on page 130, we can define a collection of single characters by enclosing them in brackets. These collections match exactly one single character. Another way to define such ranges is given in Table 9.1. They have the advantage of taking into account any variant of the local language settings or the coding system.

Table 9.1. Character classes for regular expressions

Pattern	Target
`[:alnum:]`	Any alphanumeric character 0 to 9 or A to Z or a to z.
`[:digit:]`	Only the digits 0 to 9.
`[:alpha:]`	Any alpha character A to Z or a to z.
`[:upper:]`	Any alpha character A to Z.
`[:lower:]`	Any alpha character a to z.
`[:blank:]`	Space and tabulator characters only.
`[:space:]`	Any space characters.
`[:punct:]`	Punctuation characters . , " ' ? ! ; :
`[:print:]`	Any printable character.

Sometimes you might wish to use these extended search pattern sets. However, not all programs understand them. You have to try it out.

9.2.9 Priorities

From algebra you know about the priority of arithmetic operators. The operators of multiplication/division have a larger priority (are evaluated first) than the operators of addition/substraction (which are evaluated last). There are similar rules for regular expressions. The order of priority of operators at the same parentheses level is `[ ]` (character classes), followed by `*`, `+`, `?` (closures), followed by concatenation, followed by | (alternation) and finally followed by the newline character. Parentheses have the highest priority. Thus, if you are in doubt, just use parentheses.

9.2.10 egrep Options

egrep offers a number of options that help you to fit the output to your needs.

- -v Print all lines that do not match the search pattern.
- -n Print the matched lines and corresponding line numbers.
- -c Print only the number of matches.
- -i Ignore the case of the input file. Thus the search pattern will match lower and uppercase.
- -e Protect patterns beginning with the dash (-) character.

Especially the last option can be very useful. DNA sequences are often saved in lower- or uppercase. Thus, it is not desired to distinguish between lower- and uppercase characters. You can tell egrep to ignore the case with the option -i (ignore).

9.3 Regular Expressions and vim

As stated in the beginning, regular expressions are understood by a broad range of programs. Among these is the text editor vi. In Section 6.3 on page 72 you learned how to work with this editor. The advanced version of vi, i.e. vim, allows you to highlight text. As you will see soon, this is very useful for training regular expressions.
You remember that there is a *command mode* and an *input mode*. Open a new text document, activate the input mode (i) and enter the following line of text "Animals and plants both have mitochondria. In addition, plants possess chloroplasts." Now go back to the command mode (Esc) and type ":set hls". This activates the *highlight search* mode. Now hit the / key (on my computer I have to press Shift+7). At the bottom you will see the "/" character. vi is now ready to accept a search pattern. Let us first search for a word. Enter "plants" and hit Enter. You will see that the word *plants* is highlighted. Now press / again and search for the following search pattern: "*pl.\{3\}s*". Note, it depends on your system, whether you have to escape the braces or not – I have to! After pressing Enter you will see text fragments highlighted; they all match our pattern. Now, move the cursor to the end of line. Enter the input mode and type the sentence "At many places people do research." You will see that word *places* is immediately highlighted. The last active search pattern is still active and queries the text in the background. This feature makes vi a good tool for studying and testing regular expressions. Another advantage is that vi memorizes previously used search pattern in the same way as the shell memorizes the past commands. After pressing / you can use the arrow keys ↑ and ↓ in order to scroll through the past search patterns.

Exercises

Hey folks – I do not have to tell you any more that you *must* practice. In the previous section you learned some basics of programming. In this chapter you learned the basics of pattern matching. In the coming sections we are going to fuse our knowledge. However, that is only fun when you are well prepared!

9.1. Go carefully through all the examples in this section and play around. Modify the code and observe the changes.

9.2. Find a regular expression that matches a line with exactly three space-separated fields (words).

9.3. Find a search pattern that matches a negative integer.

9.4. Design a search pattern that matches any decimal number (positive or negative) surrounded by spaces.

9.5. Find a search pattern that matches a nucleotide sequence which begins with the start codon ATG and ends with the stop codon TAA. The sequence should be at least 20 nucleotides long.

9.6. Match all lines that contain the word "hydrogenase" but omit all lines which contain the word "dehydrogenase".

9.7. A certain class of introns can be recognized by their sequence. The consensus sequences is "*GT...TACTAAC...AG*". The three dots represent an unknown number of nucleotides. Write a search pattern to match these targets.

9.8. A certain group of proteins has the following consensus sequence:
G(R/T)VQGVGFRx13G(D/W)V(C/N)Nx3G, where *(R/T)* means either R or T and *x13* stands for a sequence of 13 unspecified amino acids. The letters represent individual amino acids (one-letter amino acid code). Design a regular expression matching this target.

9.9. List all files in your home directory that are readable by all users. Do the same for all files in your home directory and all subdirectories.

10

Sed

sed (stream **ed**itor) is a non-interactive, line-oriented, stream editor. What does that mean?

Non-interactive means that the editor takes its editing commands from either the command line or a file. Once started, it runs through the whole text file that you wish to edit. That sounds worse than it is. sed does not really edit and change the original text file itself but prints the editing result to the standard output. If you want to keep the result, you just pipe it into a file.

Line-oriented means that sed treats the input file line by line. A line ends with (and a new line begins after) the (invisible) newline character. This meta character is only visible as a line break when you look at the text file with, for example, cat or vi. This has the disadvantage that, as with egrep, you cannot edit text that is spanning multiple lines. If you would like to change the words "*gene expression*" to "*transcription*" you can only do that when "*gene expression*" sits in one single line. If the end of one line contained the word "*gene*" and the beginning of the following line contained the word "*expression*", this would not be recognized. It is, however, possible to delete, insert and change multiple lines.

Stream-oriented means that sed "swallows" a whole file at once. We say: the file is treated like a stream. Imagine a text file as a pile of sheets of paper. Each sheet represents one line of the text file. Now you take one sheet, edit it according to given rules and put it aside. This you do for all sheets – you cannot stop until you have edited the last sheet. Thus, you edit a stream of sheets – sed is editing a stream of lines.

sed is a very old Unix tool. It has its roots in ed, a *line editor*. ed, however, has been completely replaced by editors like vi. In the next section we will learn why sed survived.

10.1 When to Use sed?

Why should I learn how to use a simple non-interactive editor when I know how to use vi? – you might ask. Indeed, this is a valid question in times when everybody talks about economics. Is it economic to learn sed? Yes! If you want to bake a cake "from scratch", you can either mix the dough quickly by hand, or set up the Braun super mixer for the same job. sed is the "handy" way. It is quick and has very little memory requirements. This is an advantage if you work with large files (some Megabytes). sed loads one line into the computer's memory (RAM), edits the line and prints the output either onto the screen or into a file; and sed is damn fast. I know of no other editor that can compete with sed. sed allows for very advanced editing commands. However, sed is only comfortable to use with small editing tasks. For more advanced tasks one might prefer to use awk or perl.
What are typical tasks for sed? Text substitutions! The most important thing one can do with sed is the substitution of defined text patterns with other text. Here are some examples: removing *HTML tags* from files, changing the *decimal markers* from commas to points, changing words like *colour* to *color* throughout a text document, removing *comment lines* from shell scripts, or *formatting* text files are only a small number of examples of what can be done with sed.
For repeatedly occurring tasks you can write *sed scripts* or include *sed* in shell scripts (see Program 25 on page 125 in Section 8.9.4 on page 125).

10.2 Getting Started

How is sed used? The basic syntax is quite straightforward. What makes sed scripts sometimes look complicated are the regular expressions which find the text to edit and the shortcuts used for the commands.
Let us start with a small sed editing command.

Terminal 57: Substitute Command

```
1 $ cat>test.file.txt
2 I think most spring flowers bloom white.
3 Is that caused by gene regulation?
4 $ sed 's/most/few/' test.file.txt
5 I think few spring flowers bloom white.
6 Is that caused by gene regulation?
7 $  sed -n 's/most/few/p' test.file.txt
8 I think few spring flowers bloom white.
9 $
```

In Terminal 57 we first create a short text file using cat (see Sect. 6.1 on page 64). In line 4 you find our first sed editing command. The command itself is enclosed in single quotes: s/most/few/. It tells sed to substitute the first occurrence of *most* with *few*. The slash functions as a delimiter. Altogether, there are four parts to this substitution command:

`s`	Substitute command
`/../../`	Slashes as delimiter
`most`	Regular expression pattern string
`few`	Replacement string

We will look at the substitution command in more detail in Section 10.5.1 on page 149. The editing command is enclosed in single quotes in order to avoid that the shell interprets meta characters. After executing the `sed` command in line 4 of Terminal 57 on the facing page the output is immediately printed onto the screen. Note: The input file *test.file.txt* remains unchanged.
Now you know 90% of `sed`. This is no joke. Most people use `sed` for substituting one pattern by another; but be aware that the last 10% are more difficult to achieve than the initial 90% – always! In line 7 of Terminal 57 on the preceding page `sed` is started with the option `-n`. In this case `sed` works silently and prints out text only if we tell it to do so. We do that with the command `p` (print). Each line that matches the pattern *most* is edited and printed, all other lines are neither edited nor printed onto the screen. The next terminal shows you that the single character `p` is a command.

Terminal 58: Print Command

```
1  $ sed -n 'p' test.file.txt
2  I think most spring flowers bloom white.
3  Is that caused by gene regulation?
4  $
```

In the example in Terminal 58 the option `-n` tells `sed` not to print any line from the input file unless it is explicitly stated by the command `p`.
Sometimes one likes to apply more than one editing command. That can be achieved with the option `-e`.

Terminal 59: Several Editing Commands

```
1  $ sed -e 's/most/few/' -e 's/few/most/' -e 's/white/red/'
2        test.file.txt
3  I think most spring flowers bloom red.
4  Is that caused by gene regulation?
5  $
```

All commands must be preceded by the option `-e`. The example in Terminal 59 shows how `sed` works through the editing commands. After the first line of text has been read by `sed`, the first editing command (the one most to the left) is executed. Then the other editing commands are executed one by one. This is the reason why the word *most* remains unchanged: first it is substituted by *few*, then *few* is substituted by *most*. With more and longer editing commands that looks quite ugly. You are better off writing the commands into a file and telling sed with the option -f that the editing commands are in a file.

Terminal 60: Editing Command File

```
1 $ cat>edit.sed
2 s/most/few/
3 s/few/most/
4 s/white/red/
5 $ sed -f edit.sed test.file.txt
6 I think most spring flowers bloom red.
7 Is that caused by gene regulation?
8 $
```

In Terminal 60 we create a file named *edit.sed* and enter the desired commands. Note: Now there are no single quotes or e-options! However, each command must reside in its own line.
A good way to test small sed editing commands is piping a line of text with echo to sed.

Terminal 61: echo and sed

```
1 $ echo "Darwin meets Newton" | sed 's/meets/never met/'
2 Darwin never met Newton
3 $
```

In Terminal 61 the text "*Darwin meets Newton*" is piped to and edited by sed. With this simple construction one can nicely test small sed editing scripts.
Now you know 95% of sed. Not too bad, is it? Well, the last 5% are usually even worse to grep than the last 10%. Joking apart, after this first introduction let us now take a closer look at sed.

10.3 How sed Works

In order to develop sed scripts, it is important to understand how sed works internally. The main thing one must consider is: what does sed do to a line of text? Well, first the line is copied from the input file and saved in the *pattern space*. This is sed's working memory. Then editing commands are executed on the text in the pattern space. Finally, the edited pattern space is sent to the standard output, usually the screen. Furthermore, it is possible to save a line in the *hold space*. Okay, let us examine these "spaces" a bit more closely.

10.3.1 Pattern Space

The pattern space is a kind of working memory where a single text line (also called record) is held, while the editing commands are applied. Initially, the pattern space contains a copy of the first line from the input file. If several editing commands are given, they will be applied *one after another* to the text in the pattern space. This is why we can change back a previous change as shown in Terminal 59 on page 143 (most → few → most). When all the instructions have been applied, the current line is moved to the standard output and the next line from the input file is copied into the pattern space. Then all editing commands are applied to that line and so forth.

10.3.2 Hold Space

While the pattern space is a memory that contains the current input line, the hold space is a temporary memory. In fact, you can regard it as the memory key of a calculator. You can put a copy from the pattern space into the hold space and recall it later. A group of commands allow you to move text lines between the pattern and the hold space:

`h`	*hold-O*	Overwrites the hold space with the contents of the pattern space.
`H`	*hold-A*	Appends a newline character followed by the pattern space to the hold space.
`g`	*get-O*	Overwrites the pattern space with contents of hold space.
`G`	*get-A*	Appends a newline character followed by the hold space to the pattern space.
`x`	*exchange*	Swaps the contents of the hold space and the pattern space.

You might like to play around with the hold space once you have gotten accustomed to `sed`. In the example in Section 10.6.3 on page 159, the hold space is used in order to reverse the lines of a file.

10.4 `sed` Syntax

The basic way to initiate `sed` is

```
sed 'EditCommand' Input.File
```

This line causes `sed` to edit the *Input.File* line by line according to the job specified by *EditCommand*. As shown before, if you want to apply many edit commands sequentially, you have to apply the option `-e`:

```
sed -e 'Edit.Command.1' -e 'Edit.Command.2' Input.File
```

Sometimes you wish to save your edit commands in a file. If you do this, `sed` can execute the commands from the file (which we would call a script file). Assume the script file's name is *EditCommand.File*. Then you call `sed` with

```
sed -f EditCommand.File Input.File
```

The option `-f` tells `sed` to read the editing commands from the file stated next. Instead of reading an input file, `sed` can also read the standard input. This can be useful if you want to pipe the output of one command to `sed`. We have already seen one example in conjunction with the `echo` command in Terminal 61 on page 144. A more applied example with `ls` is shown in Terminal 64 on page 150.

10.4.1 Addresses

As I said above, `sed` applies its editing commands to each line of the input file. With addresses you can restrict the range to which editing commands are applied. The complete syntax for `sed` becomes

```
sed 'AddressEditCommand' Input.File
```

Note that there is no space between the address and the editing command. There are several ways to describe an address or even a range:

`a`	*a* must be an integer describing a line number.
`$`	The dollar character symbolizes the last line.
`/re/`	*re* stands for a regular expression, which must be enclosed in slashes. Each line containing a match for *re* is chosen.
`a,b`	Describes the range of lines from line *a* to line *b*. Note that *a* and *b* could also be regular expressions or $. All kinds of combinations are possible.
`x!`	The exclamation mark negates the address range. *x* can be any of the addresses above.

Let us take a look at some examples. We do not edit anything, but just print the text lines to which the addresses match. Therefore, we run `sed` with the option `-n`. `sed`'s command to print a line of text onto the screen is `p` (see Sect. 10.5.6 on page 155). As an example file let us use the protein structure file *structure.pdb* from Section 9.1 on page 128. In order to display the whole file content we use

```
sed -n 'p' structure.pdb
```

Do you recognize that this command resembles "`cat structure.pdb`"? What would happen if we omitted the option `-n`? Then all lines would be printed twice! Without option `-n` `sed` prints every line by default *and* additionally prints all lines it is told to by the print command `p` – ergo, each line is printed twice.

Terminal 62: Addresses

```
1  $ sed -n '1p' structure.pdb
2  HEADER   Hydrogenase                              23-Mar-99    1CFZ
3  $ sed -n '2p' structure.pdb
4  COMPND   Hydrogenase Maturating Endopeptidase Hybd From
```

```
5  $ sed -n '1,3p' structure.pdb
6  HEADER  Hydrogenase                          23-Mar-99   1CFZ
7  COMPND  Hydrogenase Maturating Endopeptidase Hybd From
8  SOURCE  ORGANISM_SCIENTIFIC: Escherichia coli
9  $ sed -n '/ATOM 7/,$p' structure.pdb
10 ATOM 7  O   MET A 1 47.875  25.011  52.020  1.00 54.99 O
11 ATOM 8  SD  MET A 1 49.731  23.948  47.163  1.00 77.15 S
12
13 $ sed -n '/HELIX/p' structure.pdb
14 HELIX  1 hel ILE A   18  GLN A   28
15 HELIX  2 hel PRO A   92  THR A  107
16 HELIX  3 hel ILE A  138  SER A  152
17 $ sed -n '/HEADER/,/AUTHOR/p' structure.pdb
18 HEADER  Hydrogenase                          23-Mar-99   1CFZ
19 COMPND  Hydrogenase Maturating Endopeptidase Hybd From
20 SOURCE  ORGANISM_SCIENTIFIC: Escherichia coli
21 AUTHOR  Fritsche, Paschos, Beisel, Boeck & Huber
22 $ sed -n '/HEADER\|AUTHOR/p' structure.pdb
23 HEADER  Hydrogenase                          23-Mar-99   1CFZ
24 AUTHOR  Fritsche, Paschos, Beisel, Boeck & Huber
```

Terminal 62 gives you a couple of examples how to extract certain lines from the text file *structure.pdb*. Lines 1 to 8 show examples how to select and display certain lines or line ranges. The address used in line 9 makes `sed` print all lines from the first occurrence of "*ATOM 7*" to the last line (which is, in our case, an empty line). The address `/HELIX/` in line 13 instructs `sed` to print all lines containing the word *HELIX*. In line 22 `sed` selects all lines containing either *HEADER* or *AUTHOR*. Note that the logical *or* represented by the vertical bar needs to be escaped (\|).
It is important that you understand that the address *selects* the lines of text that are treated by `sed`. In Terminal 62 the treatment consisted simply of printing. The treatment could as well have been an editing statement. For example,

```
sed '/^ATOM/s/.*/---deleted---/' structure.pdb
```

would select all lines starting with *ATOM* (`/^ATOM/`) for editing. The editing instruction (`s/.*/---deleted---/`) selects all the content of the selected lines (`.*`) and substitutes it with the text "*—deleted—*". Try it out!

10.4.2 sed and Regular Expressions

Of course, `sed` works fine with regular expressions. However, in contrast to the meta characters we used with `egrep` in Chapter 9 on page 127, there are some syntactical differences: Parentheses "( )", braces "{ }" and the vertical bar "|" must be escaped with a backslash! You remember that parentheses group commands (see Sect. 9.2.3 on page 134) and save the target pattern for back referencing (see Sect. 9.2.7 on page 137), whereas braces belong to the quantifiers (see Sect. 9.2.2 on page 133).

10.5 Commands

Up to now, we have already encountered some vital `sed` commands. We came across substitutions with '`s/.../.../`' and know how to explicitly print lines by the combination of the option `-n` and the command `p`. We also learned that the commands `g`, `G`, `h`, `H` and `x` manipulate the pattern space and hold space. Let us now take a tour through the most important `sed` commands and their application. For some of the following examples we are going to use a new text file, called *GeneList.txt*.

GeneList.txt

```
Energy metabolism
   Glycolysis
     slr0884: glyceraldehyde 3-phosphate dehydrogenase (gap1)
     Init: 1147034 Term: 1148098 Length (aa): 354
     slr0952: fructose-1,6-bisphosphatase (fbpII)
     Init: 2022028 Term: 2023071 Length (aa): 347
Photosynthesis and respiration
   CO2 fixation
     slr0009: ribulose bisphosphate carboxylase large (rbcL)
     Init: 2478414 Term: 2479826 Length (aa): 470
     slr0012: ribulose bisphosphate carboxylase small (rbcS)
     Init: 2480477 Term: 2480818 Length (aa): 113
   Photosystem I
     slr0737: photosystem I subunit II (psaD)
     Init: 126639 Term: 127064 Length (aa): 141
     ssr0390: photosystem I subunit X (psaK)
     Init: 156391 Term: 156651 Length (aa): 86
     ssr2831: photosystem I subunit IV (psaE)
     Init: 1982049 Term: 1982273 Length (aa): 74
   Soluble electron carriers
     sll0199: plastocyanin (petE)
     Init: 2526207 Term: 2525827 Length (aa): 126
     sll0248: flavodoxin (isiB)
     Init: 1517171 Term: 1516659 Length (aa): 170
     sll1796: cytochrome c553 (petJ)
     Init: 846328 Term: 845966 Length (aa): 120
     ssl0020: ferredoxin I (petF)
     Init: 2485183 Term: 2484890 Length (aa): 97
```

This file contains information about the genome of the cyanobacterium *Synechocystis* PCC 6803. It was sequenced in 1996. The file is a very, very small exemplified assembly of annotated genes (gene names are given in the unambiguous *Cyanobase Format*, like *slr0884*, and, in parentheses, with their scientific specification), their position on the chromosome (*Init* and *Term*), the predicted length of the gene products (*aa*=amino acids) and their function. The file contains two major functional classes, *energy metabolism* and *photosynthesis and respiration*, respectively. These major classes are divided into subclasses, such as *glycolysis*, CO_2 *fixation*, *photosystem I* and *soluble electron carriers*.

10.5.1 Substitutions

Substitutions are by far the most common, and probably most useful, editing actions `sed` is used for. An example of a simple substitution is the transformation of all decimal markers from commas to points. The basic syntax of the substitution command is

```
s/pattern/replacement/flags
```

The *pattern* is a regular expression describing the target that is to be substituted by *replacement*. The *flags* can be used to modify the action of the substitution command in one or the other way. Important flags are:

- `g` Make the changes globally on all occurrences of the pattern. Normally only the first occurrence is edited.
- `n` The replacement should be made only for the *n*th occurrence of the target pattern.
- `p` In conjunction with the option `-n`, only matched and edited lines (pattern space) are printed.
- `w` *file* Same as `p`, but writes the pattern space into a file named *file*.

The flags can be used in combinations. For example, the flag `gp` would make the substitution globally on the line and print the line.
The *replacement* can make use of *back referencing* (see Sect. 9.2.7 on page 137). Furthermore, the ampersand (`&`) stands for (and is replaced by) the string that matched the complete regular expression.

Terminal 63: Back References

```
1  $ sed -n 's/^   [A-Z].*$/>>>&/p' GeneList.txt
2  >>>   Glycolysis
3  >>>   CO2 fixation
4  >>>   Photosystem I
5  >>>   Soluble electron carriers
6  $ sed -n 's/^   \([A-Z].*$\)/--- \1 ---/p' GeneList.txt
7  --- Glycolysis ---
8  --- CO2 fixation ---
9  --- Photosystem I ---
10 --- Soluble electron carriers ---
11 $
```

In Terminal 63 we see two examples of back references. In both cases we use the combination of the option `-n` and the flag `p` to display modified lines only. The target match for the search pattern "`^␣␣␣[A-Z].*$`" is any line that begins with three space characters (here symbolized by "␣"), continues with one uppercase alphanumeric character and continues with any or no character up to the end of the line. This search pattern exactly matches the subclasses in *GeneList.txt*. In line 1 of Terminal 63 the matching pattern is substituted by ">>>" plus the whole matching pattern, which is represented by the ampersand character (`&`). In line 6 we embedded only the last part of the search pattern in parentheses (note that the parentheses need to be escaped) and excluded the three spaces. The matching pattern can then be recalled with "`\1`".
In order to restrict the range of the editing action to the first 6 lines of *GeneList.txt*, line 6 of Terminal 63 could be modified to

```
sed -n '1,6s/^␣␣␣\([A-Z].*$\)/---␣\1␣---/p' GeneList.txt
```

Now let us assume we want to delete the text parts describing the position of genes on the chromosome. The correct command would be

```
sed 's/Init.*Term:␣[0-9]*␣//' GeneList.txt
```

Here, we substitute the target string with an empty string. Try it out in order to see the result.

The next example is a more practical one. In order to see the effect of the `sed` command in Terminal 64 you must have subdirectories in your home directory.

Terminal 64: Pipe to Sed

```
1  $ ls -l ~ | sed 's/^d/DIR: /' | sed 's/^[^Dt]/     /'
2  total 37472
3       rw-rw-r-- 1 Freddy Freddy     30 May  5 16:29 Datum2
4       rw-rw-r-- 1 Freddy Freddy     57 May 11 19:00 amino2s
5  DIR: rwxrwxr-x 3 Freddy Freddy   4096 May  1 20:04 blast
6  DIR: rwxrwxr-x 3 Freddy Freddy   4096 May  1 20:34 clustal
7       rw-rw-r-- 1 Freddy Freddy 102400 Apr 19 15:55 dat.tar
8  $
```

Terminal 64 shows how to use `sed` in order to format the output of the `ls` command. By applying the editing commands given in line 1, the presence of directories immediately jumps to the eye. With "`ls -l ~`" we display the content of the home directory [Remember that the tilde character (~) is the shell shortcut for the path of your home directory]. In the first `sed` editing command (`s/^d/DIR: /`) we substitute all occurrences of `d` at the beginning of a line (`^`) with "`DIR: `". The result is piped to a second instance of `sed` which introduces spaces at the beginning of each line *not* starting with a "`D`" (as in "DIR") or "`t`" (as in "total..."). With our knowledge about writing shell scripts and the `alias` function (see Sect. 7.8 on page 89), we could now create a new standard output for the `ls` command.

```
Program 26: Format ls output
1  $ cat list-dir.sh
2  #!/bin/bash
3  # save as list-DIR.sh
4  # reformats the output of ls -l
5  echo "`ls -l $1|sed 's/^d/DIR: /'|sed 's/^[^Dt]/     /'`"
```

An appropriate shell script is shown in Program 26. The shell script is executed with

`./list-DIR.sh DirName`

from the directory where you saved it. The parameter *DirName* specifies the directory of which you wish to list the content. How does the script work? We basically put the command line from Terminal 64 on the facing page into a shell script and execute the command from within the program by enclosing it in graves (see Sect. 8.5.2 on page 108). The parameter *DirName* can be accessed via the variable *$1*.

10.5.2 Transliterations

You have got a nice sequence file from a colleague. It contains important sequence data, which you want to process with a special program. The stupid thing is that the sequences are lowercase RNA sequences. However, you require uppercase DNA sequences. Hmmm – what you need is `sed`'s transliteration command "y".

```
Terminal 65: Transformation
1  $ cat>rna.seq
2  >seq-a
3  acgcguauuuagcgcaugcgaauaucgcuauuacg
4  >seq-b
5  uagcgcuauuagcgcgcuagcuaggaucgaucgcg
6  $ sed '/>/!y/acgu/ACGT/' rna.seq
7  >seq-a
8  ACGCGTATTTAGCGCATGCGAATATCGCTATTACG
9  >seq-b
10 TAGCGCTATTAGCGCGCTAGCTAGGATCGATCGCG
11 $
```

Terminal 65 shows a quick solution to the problem. In lines 1 to 5 we create our example RNA sequence file in *FASTA format*. The magic command follows in line 6. First we define the address (`/>/!`): all lines *not* containing the greater character are to be edited. The editing command invoked by "`y/acgu/ACGT/`" transliterates all "a"'s to "A"'s and so on. The transliteration command "y" does not understand words. The first character on the left

side (a) is transliterated into the first character on the right side (A) and so on. It always affects the whole line.
Now let us generate the complement of the sequences in the file *rna.seq*. The correct comment is

```
sed '/>/!y/acgu/ugca/' rna.seq
```

Could you achieve the same result with the substitution command from Section 10.5.1 on page 149? If you think so and found a solution, please email it to me.

10.5.3 Deletions

We have used already the substitution command to delete some text. Now we will delete a whole line. Assume you want to keep only the classes and subclasses but not the gene information in the *GeneList.txt* file. This means we need to delete these lines. The corresponding command is `d` (delete), the syntax is

```
addressd
```

The `d` immediately follows the address. The address can be specified as described in Section 10.4.1 on page 146. If no address is supplied, all lines will be deleted – that sounds not too clever, doesn't it? The important thing to remember is that the whole line matching the address is deleted, not just the match itself. Let us come back to our task: extracting the classes and subclasses of *GeneList.txt*.

Terminal 66: Deletions

```
1  $ sed '/     /d' GeneList.txt
2  Energy metabolism
3     Glycolysis
4  Photosynthesis and respiration
5     CO2 fixation
6     Photosystem I
7     Soluble electron carriers
8  $
```

Since all gene information lines are indented by 5 space characters, these 5 spaces form a nice address pattern. Thus, the task is solved easily, as shown in Terminal 66.

10.5.4 Insertions and Changes

Another common editing demand is to insert, append, change or delete text lines. Note that only whole lines and no words or other fragments can be inserted, appended or changed. These commands require an address. The corresponding commands are

a	*append*	Append a text after the matching line.
i	*insert*	Insert a text before the matching line.
c	*change*	Change the matching line.

Append and *insert* can only deal with a single line address. However, for *change*, the address can define a range. In that case, the whole range is changed. In contrast to all other commands we have encountered up to now, `a`, `i` and `c` are multiple-line commands. What does this mean? Well, they require a line break. For example, the syntax for the append command `a` is

```
addressa\
text
```

As you can see, the command is followed by a backslash without any space character between the address and the command! That is true for all three commands. Then follows the text that is to be appended after all lines matching the address. To insert multiple lines of text, each successive line must end with a backslash, except the very last text line.

Terminal 67: Insertion

```
1  $ sed '/>/i\
2  > ---------' rna.seq
3  ---------
4  >seq-1
5  acgcguauuuagcgcaugcgaauaucgcuauuacg
6  ---------
7  >seq-2
8  uagcgcuauuagcgcgcuagcuaggaucgaucgcg
9  $
```

Terminal 67 gives an example of an insertion. The trick with multiple-line commands is that the shell recognizes unclosed quotes. Thus, when you press (Enter) after entering line 1 of Terminal 67 the greater character (>) appears and you are supposed to continue your input. In fact, you can enter more lines, until you type the single quote again. This we do in line 2: after our insertion text (a number of dashes) we close `sed`'s command with the closing single quote followed by the filename. In this example we use the file we created in Terminal 65 on page 151. What does the command spanning from line 1 to line 2 do? Before every line beginning with a greater character some dashes are inserted.
We can do fancy things with multiple-line commands.

Terminal 68: Deletion and Insertion

```
1  $ sed '/     /d
2  > /^[A-Z]/a\
3  > ========================
4  > /   /a\
```

```
 5  > -------------------------
 6  > ' GeneList.txt
 7  Energy metabolism
 8  =========================
 9     Glycolysis
10  -------------------------
11  Photosynthesis and respiration
12  =========================
13     CO2 fixation
14  -------------------------
15     Photosystem I
16  -------------------------
17     Soluble electron carriers
18  -------------------------
19  $
```

Terminal 68 gives you an idea of how several commands can be used with one call of sed. Line 1 contains the editing commands to delete all lines containing five successive space characters. This matches the gene information lines of our example file *GeneList.txt*. The address in line 2 matches all main classes. A line containing equal characters (=) is appended to this match. Then, to all lines containing three consecutive space characters, a line with dashes is appended. The editing action is executed after hitting (Enter) in line 6. What follows in Terminal 68 is the resulting output.
Let us finally take a quick look at the *change* command.

Terminal 69: Change

```
 1  $ sed '/Photo/,$c\
 2  > stuff deleted' GeneList.txt
 3  Energy metabolism
 4     Glycolysis
 5       slr0884: glyceraldehyde 3-P dehydrogenase (gap1)
 6       Init: 1147034 Term: 1148098 Length (aa): 354
 7       slr0952: fructose-1,6-bisphosphatase (fbpII)
 8       Init: 2022028 Term: 2023071 Length (aa): 347
 9  stuff deleted
10  $
```

In the example shown in Terminal 69 all lines between the first occurrence of "Photo" and the last line is changed to (substituted by) the text "stuff deleted".

10.5.5 sed Script Files

If you need to apply a number of editing commands several times, you might prefer to save them in a file.

Program 27: Sed File

```
# save as script1.sed
# Formatting of GeneList.txt
/     /d
/^[A-Z]/a\
=========================
/   /a\
-------------------------
```

The script file containing the commands used in Terminal 68 on the facing page is shown in Program 27. Lines 1 and 2 contain remarks that are ignored by `sed`. You would call this script by

```
sed -f script1.sed GeneList.txt
```

Maybe you are using the script very, very often. Then it makes sense to create an executable script.

Program 28: Sed Executable

```
sed '
# save as script2.sed
# Provide filename at the command line
/     /d /^[A-Z]/a\
=========================
/   /a\
-------------------------
' $*
```

Program 28 would be the result of this approach. Do not forget to make the file executable with "`chmod u+x script2.sed`". Note that all commands are enclosed in single quotes. Do you recognize the variable "$*"? Take a look at Section 8.4.4 on page 105. – It contains all command line parameters. In our case these would be the files that are to be edited – yes, you can edit several files at once. In order to execute the same task as shown in Terminal 68 on the facing page you type

```
./script2.sed GeneList.txt
```

This requires much less typing, especially if you create an *alias* for the program (see Sect. 7.8 on page 89)!

10.5.6 Printing

We have already used the printing command `p` a couple of times. Still, I will shortly mention it here. Unless the default output of `sed` is suppressed (`-n`), the print command will cause duplicate copies of the line to be printed. The print command can be very useful for debugging purposes.

Terminal 70: Debugging

```
1 $ sed '/>/p
2 > s/>/</' rna.seq
3 >seq-1
4 <seq-1
5 acgcguauuuagcgcaugcgaauaucgcuauuacg
6 >seq-2
7 <seq-2
8 uagcgcuauuagcgcgcuagcuaggaucgaucgcg
9 $
```

Terminal 70 shows how to use `p` for debugging. In line 1 the current line in the pattern space is printed. Line 2 modifies the pattern space and prints it out again. Thus, we see the line before and after editing and might detect editing errors.
A special case of printing provides the equal character (`=`). It prints the line number of the matching line.

Terminal 71: Print Line

```
1 $ sed -n '/^[A-Z]/=' GeneList.txt
2 1
3 7
4 $
```

The editing command in Terminal 71 prints the line number of lines containing major classes in *GeneList.txt*.

10.5.7 Reading and Writing Files

The read (`r`) and write (`w`) commands allow you to work directly with files. This can be very comfortable when working with script files. Both commands need a single argument: the name of the file to read from or write to, respectively. There must be a single-space character between the command and the filename. At the end of the filename, either a new line must start (in script files) or the script must end with the single quote (as in Terminal 72).

Terminal 72: Write into File

```
1 $ sed -n '/^   [A-Z].*/w output.txt' GeneList.txt
2 $ cat output.txt
3    Glycolysis
4    CO2 fixation
5    Photosystem I
6    Soluble electron carriers
7 $
```

The editing command in Terminal 72 writes the subclasses from file *GeneList.txt* into the file *output.txt*.

Terminal 73: Read a File

```
1  $ cat>input.txt
2    sub classes...
3  $ sed '/^[A-Z]/r input.txt
4  > /   /d' GeneList.txt
5  Energy metabolism
6    sub classes...
7  Photosynthesis and respiration
8    sub classes...
9  $
```

In Terminal 73 we first create an input file called *input.txt*. In line 3 the file *input.txt* is inserted after each occurrence of a main class in *GeneList.txt*. Note that there must be exactly one space character between "`r`" and the filename, and that there must not be anything behind the filename except a single quote or a new line. In line 4 all other than the main class lines are deleted.
The read command will not complain if the file it should read does not exist. The write command will create a non-existing file and overwrite an existing file. However, write commands *within* one single script will append to the target file!

10.5.8 Advanced sed

As you can imagine, there are many more possibilities for what you can do with `sed`. There are some more advanced commands that allow you to work with multiple-line pattern spaces, make use of the hold space or introduce conditional branches. However, I will not go into details here. These commands require quite some determination to master and are pretty difficult to learn. In my opinion, what you have learned up to this point is by far enough to have fun with `sed`. All the advanced stuff starts to become a pain in the neck. Therefore, we save our energy in order to learn how to use `awk` and `perl` for these more advanced editing tasks. Anyway, if you feel `sed` is just the scripting language you have been waiting for, you should read some special literature in either English [3] or German [5].

10.6 Examples

You should keep up the habit of trying out some example scripts. Do some modification and follow the changes in the output.

10.6.1 Gene Tree

The following examples reformats the content of *GeneList.txt* into a tree-like output. This facilitates easy reading of the data.

Terminal 74: Gene Tree View

```
$ sed -n 's/^\([A-z]..*\)/|--- \1/p
s/^   \([A-Z]..*\)/|   |--- \1/p
s/^     [A-Za-z]..*(\(....\))/|   |   |--- \1/p
' GeneList.txt
|--- Energy metabolism
|   |--- Glycolysis
|   |   |--- gap1
|--- Photosynthesis and respiration
|   |--- CO2 fixation
|   |   |--- rbcL
|   |   |--- rbcS
|   |--- Photosystem I
|   |   |--- psaD
|   |   |--- psaK
|   |   |--- psaE
|   |--- Soluble electron carriers
|   |   |--- petE
|   |   |--- isiB
|   |   |--- petJ
|   |   |--- petF
$
```

The sed editing commands in Terminal 74 largely resemble those in Terminal 63 on page 150. A speciality resides in line 3 of Terminal 74: here we use the back reference in order to grep the content of the parentheses, that is gene name. You might have realized that the script makes use of the -n option and p command combination. This makes it easy to exclude the lines starting with "Init...".

10.6.2 File Tree

Since we started with trees, why not writing a scripting in order to print the directory content in a tree-like fashion?

Program 29: Directory Tree

```
#!/bin/bash
# save as tree.sed
# requires path as command line parameter
if [ $# -ne 1 ]; then
  echo "Provide one directory name next time!"
  echo
else
  find $1 -print 2>/dev/null |
  sed -e 's/[^\/]*\//|--- /g' -e 's/--- |/     |/g'
fi
```

What does Program 29 on the facing page do? Well, first you should recognize that it is a shell script. In line 4 we check whether exactly one command line parameter has been supplied. If this is not the case, the message "Provide one directory name next time!" is displayed and the program stops. Note: It is always a good idea to make a script *fool-proof* and inform the user about the correct syntax when he applies the wrong syntax. Otherwise, the `find` command is invoked with the directory name provided at the command line (*$1*). The `find` option `-print` prints the full filename, followed by a new line, on the standard output. Error messages are redirected into the nirvana (2>`/dev/null`, see Section 7.5 on page 85). The standard output, however, is not printed onto the screen but redirected (|) to `sed`. In order to understand what `sed` is doing, I recommend you to look at the output of `find` and apply the search pattern "`[^\/]*\/`" to it. How? Use `vi`! In Section 6.3.5 on page 75 we saw that we can read external data into an open editing file. If you type

```
:r! find . -print
```

in the command modus (and then hit (Enter)), the output of "`find . print`" is imported into `vi`. Then you can play around with regular expressions and see what the search pattern is doing, as described in Section 9.3 on page 139.

10.6.3 Reversing Line Order

Assume you have a file with lists of parameters in different lines. For reasons I do not know, you want to reverse the order of the file content. Ahh, I remember the reason: in order to exercise!

Terminal 75: Reversing Lines

```
1  $ cat > parameterfile.txt
2  Parameter 1 = 0.233
3  Parameter 2 = 3.899
4  Parameter 3 = 2.230
5  $ sed '1!G
6  > h
7  > $!d' parameterfile.txt
8  Parameter 3 = 2.230
9  Parameter 2 = 3.899
10 Parameter 1 = 0.233
11 $
```

In lines 1 to 4 of Terminal 75 we create the assumed parameter file, named *parameterfile.txt*. The `sed` script consists of three separate commands: "`1!G`", "`h`" and "`$!d`". You see, here we make use of the hold space, as explained in Section 10.3.2 on page 145. The first command in line 5 is applied to all but the first line (`1!`), whereas the third command in line 7 is applied to all but the last line (`$!`). The second command is executed for all lines. When we execute the `sed` script on the text file *parameterfile.txt*, the first command

that is executed is h. This tells sed to copy the contents of the pattern space (the buffer that holds the current line being worked on) to the hold space (the temporary buffer). Then, the d command is executed, which deletes the current line from the pattern space. Next, line 2 is read into the pattern space and the command G is executed. G appends the contents of the hold space (the previous line) to the pattern space (the current line). The h command puts the pattern space, now holding the first two lines in reverse order, back to the hold space for safe keeping. Again, d deletes the line from the pattern space so that it is not displayed. Finally, for the last line, the same steps are repeated, except that the content of the pattern space is not deleted (due to $!)' before the 'd'). Thus, the contents of the pattern space is printed to the standard output.
Note that you could execute the script also with

```
sed '1!G;h;$!d' parameterfile.txt
```

In this case, the commands are separated by the semicolon instead of the newline character.

Exercises

Now let us see what you have learned. All exercises are based on the file *structure.pdb* shown in Section 9.1 on page 128.

10.1. Change Beisel's name to Weisel.

10.2. Delete the first three lines of the file.

10.3. Print only lines 5 through 10.

10.4. Delete lines containing the word MET.

10.5. Print all lines where the HELIX line contains the word ILE.

10.6. Append three stars to the end of lines starting with "H".

10.7. Replace the line containing SEQRES with SEQ.

10.8. Create a file with text and blank lines. Then delete all blank lines of that file.

Part IV

Programming

11

Awk

Awkay, are you ready to learn a real programming language? `awk` is a text-editing tool that was developed by Aho, Weinberger and Kernighan in the late 1970s. Since then, `awk` has been largely improved and is now a complete and powerful programming language. The appealing thing for scientists is that `awk` is easy enough to be learned quickly and powerful enough to execute most relevant data analysis and transformation tasks. Thus, `awk` must not be awkward. `awk` can be used to do calculations as well as to edit text. Things that cannot be done with `sed` can be done with `awk`. Things you cannot do with `awk` you can do with `perl`. Things you cannot do with `perl` you should ask a computer scientist to do for you.

While you work through this chapter, you will notice that features similar to those we saw in the chapter on shell programming (see Chap. 8 on page 97) reappear: variables, flow control, input-output. This is typical. All programming languages have more or less the same ingredients but different syntax. On the one hand this is pain in the neck, on the other hand it is very convenient. Because if you have learned one programming language thoroughly, then you can basically work with all programming languages. You just need to cope with the different syntax; and syntax is something you can look up in a book on the particular programming language. What programming really is about is logics. First, understand the structure of the problem you want to solve, or the task you want to execute, then transfer this structure into a program. This is the joy of programming. This said, *if* you already know how to program, *then* focus on the syntax, or *else* have fun learning the basics of a programming language in this chapter. Did you recognize the italic words in the last sentence? There, we just used the control element *if-then-else*, which is part of every programming language.

As you might have guessed, there are several different versions of `awk` available. Thus, if you encounter any problems with the exercises in this chapter, the reason might be an old `awk` version on your system. It is also possible that not `awk` but `gawk` is installed on your system. `gawk` is the freeware GNU version of awk. In that case you would have to type gawk instead of awk. You can download the newest version at *rpmseek.com*.

11.1 Getting Started

Like sed, awk treats an input text file line by line. However, every line is split into fields that are separated by a space character (default) or any other definable delimiter. Each field is assigned to a variable: the whole line is stored in *$0*, the first field in *$1*, the second field in *$2* and so on.

```
Watson and Crick are smart scientists, aren't they?
------ --- ----- --- ----- ----------- ------ -----
  $1    $2   $3   $4   $5       $6        $7     $8    =  $0
```

Via the variables you have access to each field and can modify it, do calculations or whatever you want to do. As with sed, you can search for text patterns with regular expressions and execute editing commands or other actions. Furthermore, you can write awk scripts using variables, loops and conditional statements. Finally, you can define your own functions and use them as if they were built-in functions.
As an appetizer, let us take a look at an example.

Terminal 76: awk

```
1  $ cat>enzyme.txt
2  Enzyme      Km
3  Protease    2.5
4  Hydrolase   0.4
5  ATPase      1.2
6  $ awk '$2 < 1 {print $1}' enzyme.txt
7  Hydrolase
8  $
```

In Terminal 76 we first generate a new text file called *enzyme.txt*. This file contains a table with three enzymes and their catalytic activity (*Km*). In line 6 we run an awk statement. The statement itself is enclosed in single quotes. The input file is given at the end. The statement can be divided into two parts: a *pattern* and an *action*. The action is always enclosed in braces ({ }). In our example the pattern is "$2 < 1". You can read this as: "if the value in the variable *$2* is smaller than 1, then do". The content of variable *$2* is the value of field two. Since the default field separator is the space character, field 2 corresponds to the second column of our table, the *Km value*. Only in line 4 of Terminal 76 "Hydrolase 0.4" is the Km value smaller than 1. Thus, the pattern matches this line and the action will be executed. In our example the action is "{print $1}". This can be read as: "print the content of variable *$1*". The variable *$1* contains the value of field 1. In the matching line this is "Hydrolase". Therefore, field 1 is printed in line 7 of Terminal 76 on the

facing page.
The first example gave you a first insight into the function of awk. After starting awk, the lines of the input file are read one by one. If the pattern matches, then the action will be executed. Otherwise, the next line will be read. In the following section we learn more about awk's syntax.

11.2 awk's Syntax

As stated before, awk reads an input file line by line and executes an action on lines that match a pattern. The basic syntax is

```
awk 'pattern action' InputFile
```

In our example in Terminal 76 on the preceding page the pattern was "$2 < 1". This is a so-called *relational expression* pattern. However, as you will see later, the pattern could as well be a *regular expression*, a *range* pattern or a *pattern-matching expression*. The action is always enclosed in braces. In Terminal 76 on the facing page the action was "print $1". With this action, or command, the content of the first field is printed. As you can imagine, there are many other actions that can be taken and we will discuss the most important ones later. The whole awk statement, consisting of pattern and action, must be enclosed in single quotes. As usual, the input filename follows the statement, separated by a space character. You could treat more than one file at once by giving several filenames.
awk's editing statement must consist of either a pattern, an action or both. If you omit to give a pattern, then the default pattern is employed. The *default pattern* is matching every line. Thus

```
awk '{print $0}' InputFile
```

would print every line of the file named *InputFile*. The variable *$0* represents the whole currently active line of the input file. If you omit the action, the *default action* is printing the whole line. Thus

```
awk '/.*/' InputFile
```

prints all lines of *InputFile*, as well. In this case, we use the regular expression /.*/ as pattern, which matches every line.
awk comes with a great number of options. There are two important options you must remember:

- -F"x" Determines the field separator. The default setting is the space character. If you want to use another field separator you must use this option and replace *x* with your field delimiter. *x* can consist of several characters.
- -f Tells awk to read the commands from the file given after the -f option (separated by a space character).

The *default field separator* is the space character. You can change the field separator by supplying it immediately after the -F option, enclosed in double quotes.

Terminal 77: Option -F
```
1 $ awk -F":" '/Freddy/ {print $0}' /etc/passwd
2 Freddy:x:502:502::/home/Freddy:/bin/bash
3 $ awk -F":" '/Freddy/ {print $1 " uses " $7}' /etc/passwd
4 Freddy uses /bin/bash
5 $
```

In line 1 of Terminal 77 we print the complete line of the system file */etc/passwd* that matches the regular expression /Freddy/. You remember that Freddy is a username. The field separator is set to be the colon character (:). In line 3 we use awk to print the path to the shell executable, which the user Freddy is using. Again, we set the field separator to be the colon character (-F":"). For the lines containing the pattern *Freddy* (/Freddy/) we employ the print action. Here, the content of fields 1 and 7, represented by the variables *$1* and *$7*, respectively, separated by some text that must be enclosed in double quotes ("␣uses␣"), is printed.
With awk you can write very complex programs. Usually, programs are saved in a file that can be called by the option -f.

Terminal 78: Option -f
```
1 $ cat>user-shell.awk
2 /Freddy/ {print $1 " uses " $7}
3 $ awk -F":" -f user-shell.awk /etc/passwd
4 Freddy uses /bin/bash
5 $
```

In lines 1 and 2 of Terminal 78 we write the script file *user-shell.awk*. In line 3 you see how you can call a script file with the option -f. As with shell scripts or sed script files, awk script files can be commented. Comments always start with the hash character (#).

11.3 Example File

In Terminal 22 on page 65 in Section 6.1 on page 64 we created a text file with some genomic information. At certain points in this section, we are going to recycle this file, which we called *genomes.txt*. Check if you still have the file

```
find ~ -name "genomes.txt"
```

(Remember that "~" is the shell's shortcut for the path to your home directory.) Before we use this file for calculations, we perform two modifications. First, we should erase the three last lines, second, we must remove the comma delimiter in the numbers. How about a little sed statement?

Terminal 79: Modify genomes.txt

```
1 $ sed -n '/\?/!s/,//gp' genomes.txt
2 H. sapiens (human) - 3400000000 bp - 30000 genes
3 A. thaliana (plant) - 100000000 bp - 25000 genes
4 S. cerevisiae (yeast) - 12100000 bp - 6034 genes
5 E. coli (bacteria) - 4670000 bp - 3237 genes
6 $
```

Terminal 79 gives you a little sed update. We edit all lines not containing a question mark (`/\?/!` – note that the question mark must be escaped because it is a meta character) and substitute globally the comma by nothing (`s/,//g`). Then, we print only the edited lines (option `-n` plus print command `p`). The result of the sed statement should be redirected to a file named *genomes2.txt*.

```
sed -n '/\?/!s/,//gp' genomes.txt > genomes2.txt
```

Now we have a nice file, *genomes2.txt*, and we can start to learn more details about awk.

11.4 Patterns

As we have already seen, awk statements consist of a pattern with an associated action. With awk you can use different kinds of patterns. Patterns control the execution of actions: only if the currently active line, the record, matches the pattern, will the action or command be executed. If you omit to state a pattern, then every line of the input text will be treated by the actions (default pattern). Let us take a closer look at the available pattern types.
Except for the patterns BEGIN and END, patterns can be combined with the Boolean operators || (or), && (and) and ! (not).

11.4.1 Regular Expressions

Probably the simplest patterns are regular expressions (see Chap. 9 on page 127). Regular expressions must be enclosed in slashes (`/.../`). If you append an exclamation mark (`!`) after the last slash, all records *not* matching the regular expression will be chosen. In the following example we use the file *enzyme.txt*, which we created in Terminal 76 on page 164.

Terminal 80: Regular Expressions

```
1 $ cat enzyme.txt
2 Enzyme      Km
3 Protease    2.5
4 Hydrolase   0.4
5 ATPase      1.2
```

```
6  $ awk '/2/ {print $1}' enzyme.txt
7  Protease
8  ATPase
9  $ awk '!/2/ {print $1}' enzyme.txt
10 Enzyme
11 Hydrolase
12 $
```

For clarity, we first print the content of the file *enzyme.txt*. Then, in line 6 of Terminal 80, the first field (*$1*) of all records (lines) containing the character "2" is printed. The statement in line 9 inverses the selection: the first field of all records not containing the character "2" is printed. In our example, we used the easiest possible kind of regular expressions. Of course, regular expressions as patterns can be unlimitedly complicated.

11.4.2 Pattern-Matching Expressions

Sometimes you will need to ask the question if a regular expression matches a field. Thus, you do not wish to see whether a pattern matches a record but a specified field of this record. In this case, you would use a pattern-matching expression. Pattern-matching expressions use the tilde operator (~). In the following list, the variable *$n* stands for any field variable like *$1*, *$2* or so.

`$n ~ /re/`	Is true if the field *$n* matches the regular expression *re*.
`$n !~ /re/`	Is true if the field *$n* does not match the regular expression *re*.

The regular expression must be enclosed in slashes. Let us print all lines where the first field does not fit the regular expression */ase/*.

```
Terminal 81: Pattern-Matching Expression
1  $ awk '$1 !~ /ase/' enzyme.txt
2  Enzyme       Km
3  $
```

Remember that the default action is: print the line. In Terminal 81 we omit the action and specify only a pattern-matching expression. It reads: if field 1 (*$1*) does not match (!~) the regular expression (/ase/), then the condition is true. Only if the condition is true will the whole line (record) be printed. In the next example we check which users have set the bash shell as their default shell. This information can be read from the system file */etc/passwd*.

```
Terminal 82: Bash Users
1  $ awk -F":" '$7 ~ /bash/ {print $1}' /etc/passwd
2  root
3  rpm
4  postgres
```

```
5  mysql
6  rw
7  guest
8  Freddy
9  $
```

The `awk` statement in line 1 of Terminal 82 checks whether the 7th field of the */etc/passwd* file contains the regular expression "`bash`", which must be enclosed in slashes. For all matching records, the first field, that is the username, will be printed. Note that the field separator is set to the colon (`:`) character. Do you prefer to have the output ordered alphabetically? You should know how to do this! How about

```
awk -F":" '$7 ~ /bash/ {print $1}' /etc/passwd |sort
```

You just pipe the output of `awk` to `sort`.

11.4.3 Relational Character Expressions

It is often useful to check if a character string (a row of characters) in a certain field fulfils specified conditions. You might, for example, want to check whether a certain string is contained within another string. In these cases, you would use relational character expressions. The following list gives you an overview of the available expressions. In all cases the variable *$n* stands for any field variable like *$1* or *$2*. The character string *s* must always be enclosed in double quotes.

`$n == "s"`	Is true if the field *$n* matches exactly the string *s*.
`$n != "s"`	Is true if the field *$n* does not match the string *s*.
`$n < "s"`	Character-by-character comparison of *$n* and *s*. First the first character of each string is compared, then the second character and so on. The result is true if *$n* is lexicographically smaller than *s*: "flag < fly" and "abc < abcd".
`$n <= "s"`	As above; however, the result is true if *$n* is lexicographically smaller than or equal to *s*
`$n > "s"`	Character-by-character comparison of *$n* and *s*. First the first character of each string is compared, then the second character and so on. The result is true if *$n* is lexicographically greater than *s*: "house > antenna" and "abcd > abc".
`$n >= "s"`	As above; however, the result is true if *$n* is lexicographically greater than or equal to *s*

Attention, it is very easy to type "=" instead of "=="! This would not cause an error message; however, the output of the `awk` statement would not be what you want it to be. With "=" you would assign a variable. It is also worthwhile to note that uppercase characters are lexicographically less than lowercase characters and number characters are less than alphabetic characters.

$$\texttt{numbers} < \texttt{uppercase} < \texttt{lowercase}$$

Thus, "Apple" is less than "apple" or, in other words,

$$\texttt{Apple} < \texttt{apple}$$

is true. I guess we must not discuss that "Protein" is greater than "DNA", at least from a lexicographical point of view. More confusion is generated by the comparison of words starting with the same character. Let us now take a look at the results of relational character expressions in such cases.

Terminal 83: Character Relations

```
1  $ cat>chars.txt
2  Apple
3  apple
4  a
5  A
6  $ awk '$0 < "a"' chars.txt
7  Apple
8  A
9  $ awk '$0 < "A"' chars.txt
10 $ awk '$0 <= "A"' chars.txt
11 A
12 $ awk '$0 >= "A"' chars.txt
13 Apple
14 apple
15 a
16 A
17 $ awk '$0 > "A"' chars.txt
18 Apple
19 apple
20 a
21 $ awk '$0 > "a"' chars.txt
22 apple
23 $
```

I must admit that the use of the term "apple" in Terminal 83 does not quite give the impression that we are learning a programming language. Anyway, we first generate a text file with the words "Apple" and "apple" and the characters "A" and "a". In all `awk` statements we check the relation of the record (*$0*), that is the complete line, with the character "a" or "A". We do not define any action, thus, the default action (`print $0`) is used. You are welcome to try out more examples in order to get a feeling for relational character expressions. You should recognize that, in our example, the lexicographically lowest word is "A". On the contrary, the highest is the word "apple":

apple > a > Apple > A

Another important thing to understand is that "==" requires a perfect match of strings and not a partial match. Take a look back at Terminal 82 on page 169. There we looked for a partial string match. We checked whether the string "bash" (represented by the regular expression `/bash/`) is contained in the variable *$7*. If we instead used

```
awk -F":" '$7 == "bash" {print $1}' /etc/passwd
```

we would get no match. Because the variable *$7* must then have the value "bash", but it has the value "/bin/bash". This is not an exact match!

11.4.4 Relational Number Expressions

Similar to relation character expressions are relational number expressions, except that they compare the value of numbers. In the following list of available operators *$n* represents any field variable and *v* any numerical value.

`$n == v`	Is true if *$n* is equal to *v*.
`$n != v`	Is true if *$n* is not equal to *v*.
`$n < v`	Is true if *$n* is less than *v*.
`$n <= v`	Is true if *$n* is less than or equal to *v*.
`$n > v`	Is true if *$n* is greater than *v*.
`$n >= v`	Is true if *$n* is greater than or equal to *v*.

We have already used relational number expressions in our very first example in Terminal 76 on page 164. The use of relational number expressions is straightforward. In the above list, we have always used the variable *$n*. Of course, the expression is much more flexible. Any numerical value is allowed on the left side of the relation. In the following example we calculate the length of the field variable with the function `length( )`.

Terminal 84: Numerical Relation

```
1  $ awk 'length($1) > 6' enzyme.txt
2  Protease    2.5
3  Hydrolase   0.4
4  $
```

The statement in line 1 of Terminal 84 reads: print all records (default action) of the file *enzyme.txt*, where the length of the first field (`length($1)`) is larger than 6 characters. You will learn more about *functions* later.

11.4.5 Mixing and Conversion of Numbers and Characters

What happens if you mix in your relation numerical and alphabetical values? This is a tricky case and you should really take care here! Rule: If one side of your relation is a string, awk considers both sides to be strings. Thus, the following statement will lead to a wrong result.

Terminal 85: Mixing

```
1  $ awk '$2 > 2' enzyme.txt
2  Enzyme      Km
3  Protease    2.5
4  $
```

Line 2 in Terminal 85 is printed, even though "Km" is not a numerical value. However, the character string "Km" is larger than the numerical number "2" (see Sect. 11.4.3 on page 169). awk always tries to find the best solution. Thus, the statement

```
one=1; two=2; print (one two)+3
```

prints the numeric value 15. awk assumes that you want to concatenate the variables *one* and *two* with the statement "one two", leading to 12. Then awk assumes that you want to do a calculation and adds 3 to 12. The result is 15. If you need to force a number to be converted to a string, concatenate that number with the empty string "". A string (that contains numerical characters) can be converted to a number by adding 0 to that string: "2.5" converts to 2.5, "2e2" converts to 2000, "2.5abc" converts to 2.5 and "abc2.5" converts to 0. Thus, the solution to the problem in Terminal 85 is:

```
awk '($2+0) > 2' enzyme.txt
```

With these short examples I wanted to turn your attention to the problem of mixing letters and numbers. You should always carefully check your statements! Always!

11.4.6 Ranges

A range of records can be specified using two patterns separated by a comma.

Terminal 86: Range

```
1  $ awk '/En/,/Hy/' enzyme.txt
2  Enzyme      Km
3  Protease    2.5
4  Hydrolase   0.4
5  $
```

In the example in Terminal 86 all records between and including the first line matching the regular expression "En" and the first line matching the regular expression "Hy" are printed. Of course, pattern ranges could also contain relations.

Terminal 87: Pattern Range

```
$ awk '$2+0 > 3, $2+0 < 2' structure.pdb
SEQRES 4 A  162  ASP GLY GLY THR ALA GLY MET GLU LEU LEU
HELIX  1 hel ILE A   18  GLN A   28
ATOM 4  CE  MET A 1 51.349  24.403  47.765  1.00 71.39 C
ATOM 5  CG  MET A 1 48.708  24.106  48.629  1.00 66.70 C
ATOM 6  N   MET A 1 50.347  23.578  52.116  1.00 62.03 N
ATOM 7  O   MET A 1 47.875  25.011  52.020  1.00 54.99 O
ATOM 8  SD  MET A 1 49.731  23.948  47.163  1.00 77.15 S
$
```

In the example in Terminal 87 we use the file from Section 9.1 on page 128. Note that we add zero to the value of variable *$2*. By doing this, we ensure that we have a numeric value, even though the second field contains a text string. The example in Terminal 87 shows that the range works like a switch. Upon the first occurrence of the first pattern (here "`$2+0` > `3`"), all lines are treated by the action (here the default action: print record) until the second pattern (here "`$2+0` < `2`") matches or becomes true. As long as the "pattern switch" is turned on, all lines match. When it becomes switched off, no line matches, until it becomes turned on again. If the "off switch", which is the second pattern, is not found, all records down to the end of the file match. That happens in Terminal 87 from line 4 on.
If both range patterns are the same, the switch will be turned on and off at each record. Thus, the statement

```
awk '/AUTHOR/, /AUTHOR/' structure.pdb
```

prints only one line.

11.4.7 BEGIN and END

`BEGIN` and `END` are two special patterns in `awk`. All patterns described so far match input records in one or the other way. In contrast, `BEGIN` and `END` do not deal with the input file at all. These patterns allow for initialization and cleanup actions, respectively. Both `BEGIN` and `END` must have actions and these must be enclosed in braces. There is no default action.
The `BEGIN` pattern, or `BEGIN block`, is executed before the first line (record) of the input file is read. Likewise, the `END` block is executed after the last record has been read. Both blocks are executed only once.

Terminal 88: BEGIN and END

```
$ awk '
> BEGIN {FS="-"; print "Species in the file:"}
> {print "\t" $1}
> END {print "Job finished"}
> ' genomes2.txt
Species in the file:
```

```
 7             H. sapiens (human)
 8             A. thaliana (plant)
 9             S. cerevisiae (yeast)
10             E. coli (bacteria)
11     Job finished
12     $
```

In Terminal 88 we work with the file *genomes2.txt*, which we generated in Terminal 79 on page 167. Our `awk` statement spans several lines. You might remember from `sed` that the shell recognizes if opened single quotes are closed when you press (Enter). We use this feature in order to enter our `awk` program from lines 1 to 5 in Terminal 88. The `BEGIN` block contains two commands: the variable *FS* is set to the value "`-`" and some text is printed. As you will see later, the variable *FS* contains the field separator. The default value of *FS* is the space character. As you can see, we can assign the field separator either with the option `-F` (see Terminal 77 on page 166) or, as shown here, by assigning the variable *FS* in the `BEGIN` block. Note that both commands are separated by a semicolon. Alternatively, you could write both commands on two separate lines. Line 3 of Terminal 88 contains an `awk` command without pattern. Thus, the command will be applied to all lines (default pattern). The command states that "`\t`" and the content of field 1 are to be printed. As you will see later, `\t` is the tabulator meta character. Finally, the `END` block in line 4 prints a message that the job has been finished.
The special patterns `BEGIN` and `END` cannot be used in ranges or with any operators. However, an `awk` program can have multiple `BEGIN` and `END` blocks. They will be executed in the order in which they appear.

11.5 Variables

We have learned already in the section about shell scripts how to work with variables (see Sect. 8.3 on page 100). We saw that a variable stands for something else. They are a way to store a value at one point in your program and recall this value later. In `awk`, we create a variable by assigning a value to it. This value can be either a text string or a numerical. In contrast to the shell, variables must not be preceded by a dollar character. Variable names must start with an alphabetical character and can then contain any character (including digits). As everything in Linux, variable names are case-sensitive!

11.5.1 Assignment Operators

There are several ways how to assign a value to a variable in `awk`. The most common commands (assignment operators) are stated below. In the list x represents a variable, y a number or text.

`x = y`	The most common command to assign a value to a variable. y can be either a text or a digit or the result of a command.
`x += y`	The number y is added to the value of variable x. The result is stored in x. This is the same as `x=x+y`.
`x -= y`	The number y is subtracted from the value of variable x. The result is stored in x. This is the same as `x=x-y`.
`x *= y`	The number y is multiplied with the value of variable x. The result is stored in x. This is the same as `x=x*y`.
`x /= y`	The value of variable x is divided by y. The result is stored in x. This is the same as `x=x/y`.
`x ^= y`	The value of variable x is raised to the y power. The result is stored in x. This is the same as `x=x^y`.

Let us take a look at a little example on how to assign variables. Again, we work with the file *genomes2.txt*.

Terminal 89: Variables

```
1 $ awk '
2 > {species=species $1 $2 "   "
3 > genes+=$8}
4 > END{print species "\n have " genes " genes altogether"}
5 > ' genomes2.txt
6 H.sapiens   A.thaliana   S.cerevisiae   E.coli
7  have 64271 genes altogether
8 $
```

In Terminal 89 we use a multiple-line command spanning from line 1 to line 5. In line 2 we define the variable *species* and assign to it the value of itself (initially the variable is empty), plus the content of field 1 (*$1*), plus the content of field 2 (*$2*), plus 3 spaces ("␣␣␣"). Note that although *species*, *$1*, *$2* and "␣␣␣" are separated by spaces, these spaces are not saved in the variable *species*. In order to generate space characters, they must be stated within double quotes because spaces are normal text characters. In line 3, the content of field 8 of the file *genomes2.txt* is added to the content of the variable *genes*. Initially, the variable is empty (0). Since field 8 of the file *genomes2.txt* contains the number of genes, the variable *genes* contains the sum of all genes, the column sum. In line 4 of Terminal 89 `awk` is instructed to print the content of the variables *species* and *genes* together with some text that must be stated in double quotes. The special character "\n" represents the newline character. This means, not \n will be printed, but a new line will be started instead.

11.5.2 Increment and Decrement

Interesting and commonly used variable manipulation tools are the increment (`++`) and decrement (`--`) operators.

`x++` First the value of variable x is returned. Then the value of x is increased by 1: "`x=1; print x++`" → 1.

`++x` First the value of variable x is increased by 1. Then the value of x is returned: "`x=1; print ++x`" → 2.

`x--` First the value of variable x is returned. Then the value of x is decreased by 1: "`x=1; print x--`" → 1.

`--x` First the value of variable x is decreased by 1. Then the value of x is returned: "`x=1; print --x`" → 0.

As described in the list above, the increment (`++`) and decrement (`--`) operators increase or decrease the value of a variable by 1, respectively. In fact, you could do the same thing with an assignment operator described in the previous section. Thus, the pre-increment `++x` is equivalent to `x+=1` and the post-increment `x++` is equivalent to `(x+=1)-1`. The same holds for `--`: the pre-decrement `--x` is equivalent to `x-=1` and the post-decrement `x--` is equivalent to `(x-=1)-1`. The increment and decrement operators are nice, easily readable shortcuts for the assignment operators. As you might have figured out, the difference between the pre- and the post-increment (or decrement) is the return value of the expression. The pre-operator first performs the increment or decrement and then returns a value, whereas the post-operator first returns a value and then increments or decrements.

Terminal 90: Increment and Decrement

```
1  $ awk 'BEGIN{
2  > x=1; y=1; print "x="x, "y="y
3  > print "x++="x++, "++y="++y, "x="x, "y="y}'
4  x=1 y=1
5  x++=1 ++y=2 x=2 y=2
6  $
```

Terminal 90 shows you how the pre- and post-increment operator performs in real life. Both increase the value of the variables x and y by 1. However, the return values of `x++` and `++y` are different.
Terminal 90 demonstrates also the use of the `BEGIN` block for small experimental scripts. The complete `awk` script is written within the `BEGIN` block. Thus, we do not need any input file. This is a convenient way to write small scripts that demonstrate only the function of something without referring to any input file.

11.5.3 Predefined Variables

`awk` comes with a number of predefined variables that can be used and modified. While some variables provide you with valuable information about the input file or record, others allow you to adapt the behaviour of `awk` to your own needs.

Positional Variables

We have worked already intensively with positional variables. They are the only variables in `awk` that are preceded by the dollar character. The positional variable *$0* contains the whole active record, while *$1*, *$2*, *$3* and so on, contain the value of the different fields of the record.

Field Separator and Co

There are a number of very helpful variables which help you to define what structures `awk` interprets as records and fields. The best way to assign a value to any of the variables in the following list is to use a `BEGIN` block.

`FS` Field Separator. The value of *FS* is the input field separator used to split a record into the positional variables *$1*, *$2* and so on. The default value is a single-space character. However, any sequence of spaces or tabs is recognized as a single field-separator in the default setting. Furthermore, leading spaces or tabulators are ignored. You can also assign a regular expression to *FS*.

`RS` Record Separator. The value of *RS* defines a line of text. The default setting is "`\n`", which is the newline character. However, you could also assign different values to *RS*.

`OFS` Output Field Separator. This variable defines which sign is used to separate fields delimited by commas in `awk`'s `print` command. The default setting is a single-space character.

`ORS` Output Record Separator. The output record separator defines the end of each `print` command. The default setting is the newline character "`\n`".

`FIELDWIDTHS` A space-separated list of field widths to use for splitting up the record. Assigning a value to *FIELDWIDTHS* (in a `BEGIN` block) overrides the use of the variable *FS* for field splitting.

Quite commonly used is the variable *FS*. However, especially in cases where the output of scientific software is analyzed, the *FIELDWIDTHS* variable can be interesting. Let us take a look at the general use of these variables.

Terminal 91: Fieldwidth

```
1  $ awk 'BEGIN{FIELDWIDTHS="3 4 3"}
2  {print $0":\n", $1, $2, $3, $4}' enzyme.txt
3  Enzyme      Km:
4   Enz yme
5  Protease    2.5:
6   Pro teas e
7  Hydrolase   0.4:
8   Hyd rola se
9  ATPase      1.2:
10  ATP ase
11 $
```

Terminal 91 gives you an impression of how to work with the variable *FIELDWIDTHS*. In line 1 we assign the value "3 4 3" to it. This tells awk to split each new input line (record) into three fields. The first field is 3, the second 4 and the third again 3 characters long. The rest of the record is omitted. Thus, the variable *$4* is empty. Note, however, that the variable *$0* still contains the whole record. *$0* is not affected by any field splitting.
Now, let us take a look at what we can do with the output field separator.

Terminal 92: Output Field Separator

```
1  $ awk 'BEGIN{OFS="---"; print "Hello", "World"}'
2  Hello---World
3  $ awk 'BEGIN{OFS="---"; ORS="<<<"; print "One", "Two"}'
4  One---Two<<<$
```

In Terminal 92 we use only a BEGIN block to demonstrate the function of the *OFS* and *ORS* variables. As you can see, the assigned values are used to format the output of the print command. Note that the output field separator comes into effect only when commas are used with the print command. Note also that the input prompt ($) follows immediately after the output of the awk script. That is because we changed the default output field separator, the newline character (\n), to something stupid like "<<<".

File and Line Information

Some variables provide you with valuable information about the file you are just working with.

NF Number of Fields. This variable returns the number of fields of the current record (line of text).

NR Number of Records. This variable returns the number of records that have been read from the input file(s). Note that when you "digest" several files with `awk`, the number of processed records is not reset when the new file is processed!

FNR File Number of Records. This variable returns the number of records that have been read from the currently active file. In contrast to *NR*, *FNR* is reset when a new record from a new file is read!

FILENAME Filename. This variable returns the name of the currently processed file. *FILENAME* changes each time a new file is read.

Okay, let us run a stupid small script that shows us what this is all about.

Terminal 93: File Information

```
1  $ awk 'BEGIN {print "Lines & Fields in Files:"}
2  > {print FILENAME ": \t", FNR" - "NF, "\t total:" NR}
3  > ' enzyme.txt genomes2.txt
4  Lines & Fields in Files:
5  enzyme.txt:      1 - 2   total:1
6  enzyme.txt:      2 - 2   total:2
7  enzyme.txt:      3 - 2   total:3
8  enzyme.txt:      4 - 2   total:4
9  genomes2.txt:    1 - 9   total:5
10 genomes2.txt:    2 - 9   total:6
11 genomes2.txt:    3 - 9   total:7
12 genomes2.txt:    4 - 9   total:8
13 $
```

The `awk` script in Terminal 93 makes use of all variables we encountered in the list above. In line 2 we command `awk` to read out all available information on the input file given after the script in line 3. Again, we give our output some style by formatting it with the `print` command. We have learned already that commas separate output fields and lead to the output of the variable *OFS* (output field separator). The default output is a space character. The "`\t`" represents the tabulator character. It helps us to align the output. Any text we wish to print must be enclosed in double quotes. The example in Terminal 93 shows how the variables *FILENAME*, *NR*, *NF* and *FNR* are set. Take a look at the original files in order to understand the output: *enzyme.txt* is shown in Terminal 76 on page 164 and *genomes2.txt* is shown in Terminal 79 on page 167.

Command Line Parameters

The variable *ARGV* is in fact an array (see Sect. 11.5.4 on the next page). Thus, behind the facade of *ARGV* is not just one but many entries. You can access these by using *indices*. *ARGV[0]* is the first element of the array; it always contains the value "awk". *ARGV[1]* contains the first command line parameter, *ARGV[2]* the second command line parameter and so on. The variable *ARGC* returns the number of used *ARGV* variables. Thus, *ARGV[ARGC-1]* is the last command line parameter (minus 1, because the first element is 0). Note that command line parameters need to be between the `awk` script and the input file(s), separated by space characters. The syntax is:

```
   awk  'script'  par1    par2    InputFile(s)
 ARGV[0]         ARGV[1] ARGV[2]                    ARGC=3
```

Let us see how we can use command line parameters.

Terminal 94: Command Line Arguments

```
1  $ awk 'BEGIN{item=ARGV[1]; ARGV[1]=""}
2  > $1 ~ item
3  > ' rotea enzyme.txt
4  Protease    2.5
5  $
```

In line 1 of Terminal 94 we copy the value of the command line parameter in variable *ARGV[1]* to variable *item*. Then we erase the content of *ARGV[1]*. Important: you must erase all assigned *ARGV[n]* (with n>1) because they will otherwise be regarded as input files, which are not present. In order to understand this, try out the following `awk` statement:

```
awk 'BEGIN{item=ARGV[1]} $1 ~ item' rotea enzyme.txt
```

You will receive an error message saying that the file or directory could not be found. With `ARGV[1]=""` in line 1 in Terminal 94 we erase the content of *ARGV[1]*. Thus, you have to transfer the command line parameters to other variables in the `BEGIN` block and then delete the content of all `ARGV`'s. What our small script in Terminal 94 does is read one command line parameter and use it in order to find out whether it is found in the first field of any line of the input file. If so, the line will be printed out (default action).
Let us summarize what is important to remember: a) The variable *ARGV* is an array. The first command line parameter sits in *ARGV[1]* and so forth. Transfer the content of *ARGV*s to new variables inside a `BEGIN` block. Erase the content of *ARGV*s in the `BEGIN` block. That is it.
Why would we want to use command line parameters? They are helpful when you write complex `awk` scripts that are saved as files. Similarly to shell scripts, you can then use the command line parameters to influence the behaviour of your script.

Shell Environment

When you write awk scripts you might want to access some of the shell variables, like the user's home directory. For this purpose the special variable *ENVIRON* exists. In fact, this variable is an *associative array*. You need not know exactly what that is. You will learn more about arrays in Section 11.5.4. For now it is more important to see how we have to use it. Assume you want to get the path to your home directory from within a awk script.

Terminal 95: Shell Variables

```
1  $ awk 'BEGIN{print ENVIRON["HOME"]}'
2  /home/Freddy
3  $
```

The correct command is shown in Terminal 95. The brackets indicate that the variable is an array. In the same way you can gain access to all environment variables. Take a look back into Section 8.3 on page 100 for a small list of shell variables. You will find a complete list in the manual pages for bash: "man bash".
A last hint: you cannot change environmental variables from within awk scripts. This you can only do with shell scripts.

Others

There are more awk built-in variables available that we are not going to touch here. As always, you are welcome to take a look at the manpages of awk.

11.5.4 Arrays

We now learn something about a very important type of variables: *arrays*. We worked already with the arrays *ARGV* and *ENVIRON* in the previous section. In general, an array is a table of values called *elements*. The elements of an array are identified by their *indices* (also called *keys*). In awk, indices can be either numbers or text strings. Therefore, we call the arrays also *associative arrays* or *hashes*. Internally, indices are always treated as strings. For array names the same rules apply as for variable names. In addition to the name you need to state the array index, which is always done in brackets. The following line gives an example of how to assign array elements.

```
data[1]=123; data[2]="junk"; data["input"]=23.45
```

In this example the array name is *data*. As indices we used *1*, *2* and *input*. Note that strings must be enclosed in double quotes. Arrays can also have multiple dimensions like

```
data[0,0]=1; data[0,1]=2; data[1,0]=3; data[1,1]=4
```

This is a two-dimensional array. The indices for each dimension must be separated by commas.
As usual, let us take a look at a small example.

Terminal 96: Arrays
```
1  $ echo "atgccg" | awk 'BEGIN{codon["atg"]="MET"
2  > codon["ccg"]="VAL"; FIELDWIDTHS="3 3"}
3  > {print codon[$1], codon[$2]}'
4  MET VAL
5  $
```

In Terminal 96 you find a mini DNA translator. We pipe a 6-nucleotide-long DNA sequence to an `awk` script. In the `BEGIN` block we assign 2 amino acids (MET and VAL) to the array *codon*. As indices we use the codons atg and ccg, respectively. Finally, we use the built-in variable *FIELDWIDTHS* to define how to split the input string into fields. Here, we take two times three characters, our codons. In line 3 of Terminal 96 we use the resulting field variables *$1* and *$2* as indices in order to get back to the encoded amino acids. This small example shows you how powerful associative arrays can be. But now let us take a little tour through the world of arrays.

Assigning Array Elements

Assigning a value to an array element is as easy as assigning a value to a variable:

`array[index]=value`

The index can be either a number, a text string or a variable that becomes extended. This means the variable is replaced by its value (as we saw in Terminal 96 with the positional variables *$1* and *$2*). If you use a text string it must be enclosed in double quotes.
A very powerful way to create and assign an array is the command `split`:

`split(string, array, fieldseparator)`

This function divides a string at the given field separator and stores the resulting pieces in the array named *array*. The first piece would be stored in *array*[*1*]. The command returns the number of pieces that have been generated and saved.

Terminal 97: Splitting a String
```
1  $ awk '{pieces=split($2,data,".")}
2  > {print pieces, "pieces have been saved"}
3  > {print "Integer: "data[1]"\tDecimal: "data[2]}
4  > ' enzyme.txt
5  1 pieces have been saved
6  Integer: Km  Decimal:
7  2 pieces have been saved
```

```
 8  Integer: 2   Decimal: 5
 9  2 pieces have been saved
10  Integer: 0   Decimal: 4
11  2 pieces have been saved
12  Integer: 1   Decimal: 2
13  $
```

Terminal 97 gives a nice example of the `split` command. In order to recall the content of the file *enzyme.txt* take a look at Terminal 98 on the following page. What are we doing in Terminal 97? Well, first we split the content of variable *$2* at the point character. Since we know that there is only one point in this field, the decimal delimiter, we expect to obtain two elements for the array *data*. These we print out. We now have the integer part and decimal part of the Km value separated. As you can see from line 6 of Terminal 97, only one array element is generated if the defined delimiter is not found. This is the case in the first line of *enzyme.txt*.
As with the input line field-splitting, leading and trailing spaces as well as running spaces are ignored when the field separator is set to a single-space character. If you do not provide any field separator with the `split` command, the `awk`'s variable *FS* will be used instead.

Referring to an Array Element

There are two ways in which you can refer to an array element. The most common one is to use the array's name and an index:

`array[index]`

The other way is to use the expression `in`:

`index in array`

The result of the expression is true if the element *array[index]* exists.

Scanning an Array

There is no way to check whether a certain value is contained within an array. This is a pity, but the bitter truth. Instead, `awk` provides a special command to scan through a complete array:

`for (variable in array) commands`

This `for...in...` command generates a loop through all previously used indices of the array *array*. These indices are accessible via the variable *variable*. The reality looks like this:

Terminal 98: For...In...

```
1  $ awk '/[0-9]/ {data[$1]=$2}
2  END{for (x in data) print "Index="x, "  Value="data[x]}
3  ' enzyme.txt
4  Index=Hydrolase   Value=0.4
5  Index=Protease   Value=2.5
6  Index=ATPase   Value=1.2
7  $ cat enzyme.txt
8  Enzyme       Km
9  Protease     2.5
10 Hydrolase    0.4
11 ATPase       1.2
12 $
```

The script in Terminal 98 saves all values of the first and the second field of lines with a number in the array *data*. In the `END` block we use the `for...in...` loop to recall all array elements and print out their value (`data[x]`), together with the index (`x`). In line 7 we print out the file *enzyme.txt* again. If you compare the output of the `awk` script with the file content you will recognize that the order of the output is strange. There is nothing we can do about this. The order in which the elements of the array are accessed by `awk` cannot be influenced. It depends on the internal processing of the data.

Deleting an Array or Element

In order to delete a complete array you just use the command

```
delete array
```

where *array* is the array name. If you want to delete only a certain element you can use

```
delete array[index]
```

Once an array element has been deleted, it is gone forever. There is no way to recover it.

Sorting an Array

With the command `asort` you can sort the elements of an array according to their value.

Terminal 99: Sorting

```
1  $ awk '/[0-9]/ {data[$1]=$2}
2  END{asort(data)
3  for (x in data) print "Index="x, "  Value="data[x]}
4  ' enzyme.txt
5  Index=1   Value=0.4
```

```
6 Index=2   Value=1.2
7 Index=3   Value=2.5
8 $
```

In Terminal 99 we use the `asort` command in an `END` block. As you will notice in the output, all original indices of the array *data* are irrecoverably lost! This is not quite what you want to happen. The only thing you can do is use the `asort` command as

```
asort(inarray, outarray)
```

With this construction, `awk` copies the *inarray* into *outarray* and then sorts *outarray*. In this way, only the indices of the newly generated *outarray* are destroyed, whereas your original *inarray* remains untouched.
If you want to sort the indices of an array, you have a problem – a problem that can be attacked with a little program. You will find it in Section 11.10.1 on page 210.

11.5.5 Shell Versus `awk` Variables

You should be aware of the fundamental difference between shell variables (see Sect. 8.3 on page 100) and `awk` variables. To recall shell variables you must precede them with the dollar character. When assigning a value to a shell variable, you do not precede it with the dollar character. In `awk`, you *never* use the dollar character with variables. The only exceptions are special variables like the record (*$0*) or fields (*$1, $2...*).

11.6 Scripts and Executables

Up to this point we have had many examples of scripts that spanned more than one line. This is not quite convenient, especially not when you want to reuse these scripts. Then it would be much nicer to have them stored in files. Like shell scripts and `sed` scripts you can, of course, also save `awk` scripts. The syntax to call a script file is

```
awk -f ScriptFile InputFile(s)
```

As usual, script files can be commented. Comments always begin with the hash character (`#`).

Program 30: Sort Array

```
1 # save as sort1.awk
2 # sort on value
3 /[0-9]/ {data[$1]=$2}
4 END{asort(data)
5 for (x in data) {print "Index="x, "  Value="data[x]}
6 }
```

Program 30 matches the script we used in Terminal 99 on the preceding page. You call the script with

```
awk -f sort1.awk enzyme.txt
```

Like shell scripts, you can make awk scripts executable. This is very convenient when you use the scripts a lot and want to create an alias.

Program 31: Executable Script

```
#!/bin/awk -f
# save as sort1-exe.awk
# sort on value
/[0-9]/ {data[$1]=$2}
END{asort(data)
for (x in data) {print "Index="x, "  Value="data[x]}
}
```

Program 31 gives an example of an executable awk script. It is important to remember to use the option -f in the first line. Of course, you must make the file executable with

```
chmod u+x sort1-exe.awk
```

The script is executed with

```
./sort1-exe.awk enzyme.txt
```

This is basically all you need to know in order to create awk scripts.

11.7 Decisions – Flow Control

In this section you will learn how to influence the flow of your awk programs. You should already be an expert in this – at least, if you have worked carefully through the corresponding section on shell programming (see Sect. 8.7 on page 110). Decisions, or *control statements* as one might call them, are almost the most important part of a programming language. If you look at sed, you will realize that this is a very limited tool – because it lacks solid control structures. awk's control structures belong to the world of actions, this means they must be enclosed in braces. The control statements in this section check the state of a condition, which can be either true or false. This condition is typically one of the relational expression presented in Sections 11.4.2 on page 168, 11.4.3 on page 169 and 11.4.4 on page 171. Okay, let us see how we can use control statements.

11.7.1 if...else...

If you read up to this point *then* read on, or *else* start with Chapter 11 on page 163. – This is what the `if...else` construct is doing: it checks if the first condition is true. If it is true, the following statement is executed, or else an alternative is executed. This alternative is optional.

```
{if (condition) {
   command-1(s)}
else {                     # this part is optional
   command-2(s)}           # this part is optional
}
```

The "`else {command-2(s)}`" is optional and can be omitted. Let us take a look at a little program.

Program 32: if...else

```
1  # save as if-else.awk
2  # demonstrates if-else structure
3  # use with enzyme.txt
4  {if ($2 < 2) {
5    sum_b+=$2}
6  else {
7    sum_s+=$2}
8  }
9  END{
10 print "Sum of numbers greater than or equal 2: "sum_b
11 print "sum of numbers smaller than 2          : "sum_s
12 }
```

Program 32 shows an example of how to use the "`if...else`" construction. The program is executed by

```
awk -f if-else.awk enzyme.txt
```

It simply calculates the sum of the value of all second fields of the file *enzyme.txt* that are smaller than 2 (`if ($2 < 2)`) and larger than or equal to 2 (`else`). Remember that "`sum_b += $2`" (where *sum_b* is a variable) is the same as "`sum_b = sum_b + $2`" (see Sect. 11.5.1 on page 174). The output of the program is shown in the following Terminal.

Terminal 100: if...else

```
1 $ awk -f if-else.awk enzyme.txt
2 Sum of numbers greater than or equal 2: 1.6
3 sum of numbers smaller than 2          : 2.5
4 $ cat enzyme.txt
5 Enzyme      Km
6 Protease    2.5
7 Hydrolase   0.4
```

```
8  ATPase       1.2
9  $
```

Just to remind you of the content of *enzyme.txt*, it is printed out with the `cat` command.

11.7.2 `while...`

While you read this script, concentrate! – The `while` statement does something (concentrating), while a condition is true (reading). In contrast to the "`do...while`" construction (see Sect. 11.7.3 on the next page), first the state of the condition is checked and, if the condition is true, commands are executed. The result is a loop that will be repeated until the condition becomes false. The syntax is:

```
{while (condition) {
  command(s)}
}
```

It is quite easy to end up in an endless loop with the `while` command. Keep this in mind whenever you use it!

Program 33: while

```
1  # save as while.awk
2  # demonstrates the while command
3  # reverses the order of fields
4  BEGIN{ORS=""}
5  {i=NF}
6  {while (i>0) {
7    print $i"\t"; i--}
8  }
9  {print "\n"}
```

Program 33 shows you an application which prints out the fields of each line in reverse order. This means the last field is becoming the first, the second last is becoming the second and so forth. Execution and output of the program is shown in Terminal 101.

Terminal 101: while

```
1  $ awk -f while.awk enzyme.txt
2  Km      Enzyme
3  2.5     Protease
4  0.4     Hydrolase
5  1.2     ATPase
6  $
```

In the `BEGIN` block in Program 33 we reset the output record separator variable *ORS* (see Sect. 11.5.3 on page 177). As a result, the `print` command will not generate a new line after execution. For each line of the input file the number of fields (variable *NF*) is stored in the variable *i*. In the `while` loop, first the highest field number (*$i* = *$(NF)*) is printed, followed by a tabulator character "`\t`". Next, *i* is decremented by 1 (`i--`). The `while` loop stops when *i* becomes 0. Then a newline character "`\n`" is printed and the next line (record) read from the input file.

11.7.3 `do...while...`

Do not drink alcohol *while* you learn how to use control structures. – In contrast to the `while` command (see Sect. 11.7.2 on the facing page), first one or more commands are executed and then it is checked if the condition is true. Only if the condition is still true, is execution of the commands repeated. The syntax is:

```
{do {
   command(s)}
  while (condition)
}
```

Since the `do...while` construct is very similar to the `while` command, I will not present any example here. Just remember that the command exists and that it, in contrast to `while`, executes the command(s) at least once, even if the condition is false.

11.7.4 `for...`

For each section, there is an example you should carefully study. – The `for` loop is a special form of the `while` loop. We saw already a, though unusual, example of a `for` in Section 11.5.4 on page 181. There it was used to read out the complete content of an array.
The normal `for` loop consists of three parts: *initialization*, *break condition* and *loop counter*. The syntax is:

```
{for (initialization; condition, counter){
   command(s)}
}
```

The first part, the initialization, is executed only once. Here, a counter variable is set. The condition is checked in every loop. Only when the condition is true will the commands be executed and the loop counter be executed. As loop counter one uses usually the increment (`++`) or decrement (`--`) command.
The following program prints the quadratic powers of the numbers 1 to 10.

Program 34: for

```
# save as for.awk
# demonstrates the for command
# calculates quadratic numbers
BEGIN{ORS=" "
for (i=1; i<=10; i++){
  print i**2}
print "\n"}
```

I guess the program almost explains itself. The whole script is executed in a `BEGIN` block because we do not need any input file. The output record separator variable *ORS* is set to the space character. This prevents the `print` command from generating line breaks. "`i**2`" raises the value of the variable *i* to the power of 2 (see Sect. 11.8.2 on page 195). In line 7 we print a newline character (`\n`). Execution of the program is shown in Terminal 102.

Terminal 102: for

```
$ awk -f for.awk
1 4 9 16 25 36 49 64 81 100
$
```

11.7.5 Leaving Loops

Being in a loop is fine, as long as you want to be there. However, it might be necessary to leave a `while` or `for` loop before it is officially finished by the conditional state. Of course, there are nice statements for this.

break – With the `break` command you can simply jump out of a loop. The program execution continues at the first command following the loop.

```
while (condition){
  command-L1
  command-L2
  break              >-
  command-L3}          |
command              <-
```

continue – With the `continue` command you leave the actual loop. However, in contrast to `break`, it brings you back to the beginning of the loop. Only the remaining commands in the loop are skipped and the next cycle is initiated.

```
while (condition){  <-
  command-L1           |
  command-L2           |
  continue           >-
  command-L3}
command
```

next – The next command forces awk to immediately stop processing the current record and load the next record from the input file. The next command is not only interesting in the context of loops but can be generally used in awk scripts. With the command nextfile you can even force awk to skip the current input file and load the first line of the next input file. If there is no other input file, nextfile works like exit.

exit – The exit command causes awk to stop the execution of the whole awk script. The syntax is

```
exit n
```

The number n is an optional extension. If you use it, it will be the return code of your awk script. After execution of the exit statement, the script first tries to execute the END block. However, if no END block is present, the program stops immediately. If the exit command is executed in the BEGIN block, awk does not try to jump to the END block but stops execution directly.

11.8 Actions

Until now, we have spent a lot of time on understanding patterns, variables and control structures. This is indispensable knowledge in order to work with a programming language like awk. However, the real power of awk unfolds when we learn about its actions (functions) in this and the following sections. You have learned already that awk actions are always enclosed in braces. Everything enclosed in braces are actions, which are also called functions. We have already made extensive use of the print function. When we worked with arrays in Section 11.5.4 on page 181 we used the functions asort and split. In this section we will deal with many more powerful functions. In Section 11.8.5 on page 201 you will even learn how we define our own functions.

11.8.1 Printing

The default print command is "print $0". Thus, if you use only print in your script, awk will interpret this as "print $0". The syntax of the print command is

```
print item1, item2, ...
```

The items to be printed out are separated by commas. If you want to print text, it must be enclosed in double quotes:

```
print "Text: " variable
```

Everything within double quotes will be printed as it is, whereas *variable* is assumed to be a variable. Thus, the value of the variable will be printed. Each comma is substituted by the value of the output field separator variable *OFS*. The default value is the space character "␣". Furthermore, each print

command is finished by printing the value of the output record separator variable *ORS*. The default value is the newline character "\n".
Usually, `print` prints to the standard output, that is the screen. However, you can also use *redirections* in order to print into a file.

```
awk 'print > "output.txt"' enzyme.txt
```

This command would print all lines of the file *enzyme.txt* into the file *output.txt*. If the file *output.txt* does not exist, it will be created. If it exists, it will be overwritten, unless you use >>, which appends the output to an existing file.
Now let us take a look at some special characters. They are used to print tabulators, start new lines or even let the system bell ring. The following list shows the most important escape sequences.

- `\a` You will hear an *alert* (system bell) when you print this character.
- `\b` This prints the *backspace* character. The result is that the printing cursor will be moved backwards.
- `\n` When this character is printed a *new line* will be started.
- `\t` A *horizontal tabulator* is printed.
- `\v` A *vertical tabulator* is printed.

The following Terminal gives an example of how to use escape sequences.

Terminal 103: Backspace

```
1 $ awk '{print $1"\b\b\bx"}' enzyme.txt
2 Enzxme
3 Protexse
4 Hydrolxse
5 ATPxse
6 $
```

The `awk` statement in line 1 of Terminal 103 prints out the first column of the file *enzyme.txt*. After the content of the positional variable *$1* is printed, the cursor is placed three positions back (`\b\b\b`), then the character "x" is printed. By this, the original content of the line is overwritten.

The command `printf` provides more control over the format of the output. The syntax is:

```
printf (format, item1, item2,...)
```

Here, *format* is a string, the so-called *format string*, that specifies how to output the items. Note that `printf` does not append a newline character! It prints only what the format string specifies. Both the output field separator variable *OFS* and the output record separator variable *ORS* are ignored.

The format string consists of two parts: the *format specifier* and an optional *modifier*. Table 11.1 shows a number of available format specifiers and gives examples of what they do.

Table 11.1. Function of format specifiers for `printf`

Spec.	Value	Result	Comment
%s	10.63	*10.63*	print a string
%i	10.63	*10*	print only the integer part of the number
%e	10.63	*1.063000e+01*	print a number in scientific notation
%f	10.63	*10.630000*	print a number in floating point notation
%g	10.63	*10.63*	print a number either in floating point or scientific notation, whichever uses less characters

Note that you need to give a format specifier for each item you want to print. That is illustrated in Terminal 104.

Terminal 104: Format Specifier

```
1 $ awk 'BEGIN{printf("%g\n", 10.63, 5)}'
2 10.63
3 $ awk 'BEGIN{printf("%g%g\n", 10.63, 5)}'
4 10.635
5 $ awk 'BEGIN{printf("%g - %g\n", 10.63, 5)}'
6 10.63 - 5
7 $
```

In line 1 of Terminal 104 we use the format string "%g\n". It reads: print a number in the shortest possible way, followed by a newline character. Remember that `printf` does not start a new line by itself. As you can see from the output in line 2, the second item is omitted. In line 3 we specify to print 2 numbers. They are printed one after the other. In line 5 we introduce a separator between the numbers.
With modifiers you can further specify the output format. Modifiers are specified between the % sign and the following letter in the format specifier. Now let us take a look at some available modifiers.

w Width. This is a number specifying the desired minimum width of a field. If more characters are required, the field width will be automatically expanded accordingly. By default, the output is right-aligned.

- Preceding the *with* with the minus character leads to a left alignment of the output.

␣ In numeric output, positive values will be prefixed with a space and negative values with a minus sign.

+ In numeric output, positive values will be prefixed with the plus sign and negative values with the minus sign.

.n Precision. A periodic followed by an integer gives the precision required for the output: "`%.ns`": maximum n characters are printed; "`%.ni`": minimum number of digits is n; "`%.ng`": maximum number of significant digits; "`%.ne`" or "`%.nf`": n digits to the right of the decimal point.

The following Terminal shows some examples of how the modifiers are used in conjunction with the format modifiers for strings (`%s`) and floating point numbers (`%f`).

Terminal 105: Modifiers

```
1  $ awk 'BEGIN{
2  printf("%s - %f\n", "chlorophyll", 123.456789)}'
3  chlorophyll - 123.456789
4  $ awk 'BEGIN{
5  printf("%6s - %6f\n", "chlorophyll", 123.456789)}'
6  chlorophyll - 123.456789
7  $ awk 'BEGIN{
8  printf("%15s - %15f\n", "chlorophyll", 123.456789)}'
9      chlorophyll -      123.456789
10 $ awk 'BEGIN
11 {printf("%-15s - %-15f\n", "chlorophyll", 123.456789)}'
12 chlorophyll     - 123.456789
13 $ awk 'BEGIN
14 {printf("%+15s - %+15f\n", "chlorophyll", 123.456789)}'
15     chlorophyll -     +123.456789
16 $ awk 'BEGIN{
17 printf("%.6s - %.6f\n", "chlorophyll", 123.456789)}'
18 chloro - 123.456789
19 $ awk 'BEGIN{
20 printf("%15.6s - %15.6f\n", "chlorophyll", 123.456789)}'
21          chloro -      123.456789
22 $ awk 'BEGIN{
23 printf("%-15.6s - %-15.6f\n", "chlorophyll", 123.456789)}'
24 chloro          - 123.456789
25 $
```

Terminal 105 on the facing page gives you an impression of the function of the modifiers. As you can see, they are placed between the % character and the format control letter (here, s and f). All number values without or preceding the period character (.) are *width* values, specifying the minimum width. In lines 11, 14 and 23 the width value is preceded by another formatting character. All numbers behind the period character specify the precision of the output. For details, you should consult the list above.

Finally, `awk` provides the command `sprintf`. With this function `awk` redirects the output into a variable instead of a file.

Terminal 106: Print Into Variable

```
1  $ awk 'BEGIN{
2  > out=sprintf("%s","Hello World")
3  > print out}'
4  Hello World
5  $
```

In the example in Terminal 106 we use `sprintf` in order to assign some formatted text to the variable *out*. In line 3 the content of *out* is directed onto the standard output (printed on the screen).

11.8.2 Numerical Calculations

Apart from the standard arithmetic operators for addition (`x+y`), substraction (`x-y`), division (`x/y`), multiplication (`x*y`) and raising to the power (`x**y` or `x^y`), `awk` comes with a large set of numeric functions that can be used to execute calculations. The following list gives you a complete overview of the built-in arithmetic functions.

`int(x)`	Truncates x to integer.
`log(x)`	The natural logarithm function.
`exp(x)`	The exponential function.
`sqrt(x)`	The square root function.
`sin(x)`	Returns the sine of x, which must be in radians.
`atan2(y, x)`	Returns the arctangent of y/x in radians.
`cos(x)`	Returns the cosine of x, which must be in radians.
`rand()`	Returns a random number between 0 and 1, excluding 0 and 1. With "`int(n*rand())`" you can obtain non-negative integer random number less than n.
`srand([x])`	Uses x as a new seed for the random number generator. x is optional. If no x is provided, the time of day is used. The return value is the previous seed for the random number generator.

The use of these functions is quite straightforward.

Terminal 107: Mathematics

```
1  $ awk 'BEGIN{srand()
2  print log(100), sin(0.5), int(5*rand()), 3**2}'
3  4.60517 0.479426 1 9
4  $
```

In line 1 in Terminal 107 we set a new value for random number generator. Then the results of some calculations are printed.

11.8.3 String Manipulation

Probably the most important functions are those that manipulate text. The following functions are of two different types: they either modify the target string or not. Most of them do not. All functions give one or the other type of return value: either a numeric value or a text string. As with the functions for numerical calculations in Section 11.8.2 on the page before, this list of functions is complete.

`substr(target, start, length)`
Substring. – The `substr` function returns a *length* character long substring of *target* starting at character at position *start*. If *length* is omitted, the rest of *target* is used.

Terminal 108: substr

```
1  $ awk 'BEGIN{target="atgctagctagctgc"
2  > print substr(target,7,3)
3  > print target}'
4  gct
5  atgctagctagctgc
6  $
```

Terminal 108 illustrates how to use `substr`. Here we extract and print 3 characters from the 7th position of the value of the variable *target*.

`gsub(regex, substitute, target)`
Global Substitution. – Each matching substring of the regular expression *regex* in the target string *target* is replaced by the string *substitute*. The number of substitutions is returned. If *target* is not supplied, the positional variable *$0* is used instead. An `&` character in the replacement text *substitute* is replaced with the text that was actually matched by *regex*. To search for the literal "&" type "\\&".

Terminal 109: gsub

```
1  $ awk 'BEGIN{target="atgctagctagctgc"
2  > print gsub(/g.t/,">&<",target)
3  > print target}'
4  3
```

```
5  at>gct<a>gct<a>gct<gc
6  $
```

In the example in Terminal 109, the regular expression "`g.t`", enclosed in slashes, is replaced by the matching string (`&`) plus greater and less characters. Note that the target string itself is changed and that the function returns the number of substitutions made.

`sub(regex, substitute, target)`
Substitution. – Just like `gsub()`, but only the first matching substring is replaced.

Terminal 110: sub

```
1  $ awk 'BEGIN{target="atgctagctagctgc"
2  > print sub(/g.t/,">&<",target)
3  > print target}'
4  1
5  at>gct<agctagctgc
6  $
```

As shown in Terminal 110, `sub` substitutes only the first occurrence of the regular expression. Thus, the return value can only be 1 (one substitution) or 0 (regular expression not found in target string).

`gensub(regex, substitute, how, target)`
General Substitution. – Search the target string *target* for matches of the regular expression *regex*. If *how* is a string beginning with "g" or "G" (global), then all matches of *regex* are replaced with *substitute*. Otherwise, *how* is a number indicating which match of *regex* is to be replaced. If *target* is not supplied, the positional variable *$0* is used instead. Back referencing is allowed (see Sect. 9.2.7 on page 137). Note that unlike `sub()` and `gsub()`, `gensub` returns the modified string. The original target string is not changed.

Terminal 111: gensub

```
1  $ awk 'BEGIN{target="atgctagctagctgc"
2  > print gensub(/g.t/,">&<",2,target)
3  > print target}'
4  atgcta>gct<agctgc
5  atgctagctagctgc
6  $
```

Terminal 111 gives you an example of the function `gensub`. Here, we substitute the 2nd occurrence of the regular expression in *target*. Note that the original string is not modified. Instead, the modification of the text in the variable *target* is the output of `gensub`.

`index(target, find)`
Search. – This function searches within the string *target* for the occurrence of the string *find*. The index (position from the left, given in characters) of the string *find* in the string *target* is returned (The first character in *target* is number 1). If *find* is not present, the return value is 0.

Terminal 112: index
```
$ awk 'BEGIN{target="atgctagctagctgc"
> print index(target,"gct")
> print target}'
3
atgctagctagctgc
$
```

Terminal 112 gives you an example of the function `index`. It simply returns the position of the given string.

`length(target)`
Length. – Another easy-to-learn command is `length`. As the name implies, this function returns the length of the string *target*, or the length of the positional variable *$0* if *target* is not supplied.

Terminal 113: length
```
$ awk 'BEGIN{target="atgctagctagctgc"
> print length(target)
> print target}'
15
atgctagctagctgc
$
```

Terminal 113 gives you an example of how to use `length`.

`match(target, regex, array)`
Matching. – The `match` function returns the position where the longest, left-most match by the regular expression *regex* occurs in the target string *target*. If *regex* cannot be found, 0 is returned. The built-in variables *RSTART* and *RLENGTH* are set to the index (position from the left, counted in characters) and the length of the matching substring, respectively.
If *array* is used, the element *array[0]* will be set to the matching substring of *target*. Elements *1* through *n* are filled with the portions of *target* that match the corresponding parenthesized subexpression of the regular expression *regex* (see the Sect. on back referencing, 9.2.7 on page 137).

Terminal 114: match
```
$ awk 'BEGIN{target="atgctagctagctgc"
> print match(target,/(g.(.*).t)/,array)
> print "Target: "target
> print "0: "array[0], "1: "array[1], "2: "array[2]}'
```

```
5 3
6 Target: atgctagctagctgc
7 0: gctagctagct 1: gctagctagct 2: tagctag
8 $
```

Terminal 114 shows an example of `match`. Note that the contents of *array[1]* and *array[2]* overlap by 5 characters. This is due to our nice arrangement of parentheses in the regular expression! Take a careful look at it and figure out what is going on.

`split(target, array, regex)`
Splitting. – The `split` function splits the string *target* and assigns the resulting products to the elements of an array called *array* (consult also Sect. 11.5.4 on page 181). The first element of the array is *array[1]* and not *array[0]*! The desired field separator to split *target* is given by the regular expression *regex* which, of course, could also be a string. The number of generated fields is returned. If *regex* is omitted, the built-in field separator variable *FS* will be used instead.

Terminal 115: aplit

```
1 $ awk 'BEGIN{target="atgctagctagctgc"
2 n=split(target,array,/g.t/)
3 for (i=1; i<=n; i++){printf "%s", array[i]"  "}
4 print ""}'
5 at  a  a  gc
6 $
```

In Terminal 115 we split the string saved in the variable *target* at positions defined by the regular expression "`g.t`". The resulting fields are saved in the array named *array*, whereas the number of resulting fields is saved in *n*. The `for` loop in line 3 is used to read out all elements of the array.

`asort(array, destination)`
Array Sorting. – This function returns the number of elements in the array *array*. The content of *array* is copied to the array *destination*, which then is sorted by its values (for more information see Sect. 11.5.4 on page 181). The original indices are lost. Sorting is executed according to the information given in Section 11.4.3 on page 169 (apple > a > A > Apple).

Terminal 116: asort

```
1 $ awk 'BEGIN{target="atgctagctagctgc"
2 n=split(target,array,/g.t/); asort(array)
3 for (i=1; i<=n; i++){printf "%s", array[i]"  "}
4 print ""}'
5 a  a  at  gc
6 $
```

The script in Terminal 116 is the same as in Terminal 115 on the preceding page, but extended by the `asort` function. Thus, the elements of the array are sorted according to their value, before they are printed out.

`strtonum(target)`
String To Number. – This function examines the string *target* and returns its numeric value. You will hardly need this function.

Terminal 117: strtonum
```
1 $ echo "12,3" | awk '{print strtonum($0)}'
2 12
3 $ echo "12.3" | awk '{print strtonum($0)}'
4 12.3
5 $ echo "alpha12.3" | awk '{print strtonum($0)}'
6 0
7 $
```

Terminal 117 illustrates the function of `strtonum`.

`tolower(target)`
To Lowercase. – Returns a copy of the string *target*, with all uppercase characters converted to their corresponding lowercase counterparts. Non-alphabetic characters are left unchanged. For an example take a look at its counterpart `toupper`.

`toupper(target)`
To Uppercase. – The opposite of the function `tolower()`.

Terminal 118: toupper
```
1 $ awk 'BEGIN{target="atgctagctagctgc"
2 > print toupper(target)
3 > print target}'
4 ATGCTAGCTAGCTGC
5 atgctagctagctgc
6 $
```

Well, the function `toupper` really does what it says. Terminal 118 gives you an example.

11.8.4 System Commands

In the same way that you invoke an `awk` statement from a shell script, you can invoke a shell command, or system call, from your `awk` script. The `system` function executes the command given as a string.

Terminal 119: System Call
```
1 $ awk 'BEGIN{system("pwd")}'
2 /home/Freddy
```

```
3  $ awk 'BEGIN{print system("pwd")}'
4  /home/Freddy
5  0
6  $ awk 'BEGIN{print system("pwdd")}'
7  sh: line 1: pwdd: command not found
8  127
9  $ awk 'BEGIN{print "pwd" | "/bin/bash"}'
10 /home/Freddy
11 $
```

Terminal 119 demonstrates different ways in which you could execute a system call. In line 1 we execute a simple system call with the command `pwd`. The result is printed to standard output. In line 3 we include the `print` function. Now, `awk` prints out the return value of the system call. 0 means that there was no error, in contrast to lines 6 to 8. In line 9 we redirect the output of the print command to the shell. This is a quite convenient way to execute a shell command.
In Section 11.9 on page 206 you will learn how you can use the output of a system call in an `awk` program.

11.8.5 User-Defined Functions

It is not my intention to present all available functions here. A complete list of `awk`'s functions can be found in the manpages; but what happens if your desired function is missing? Then you have to create your own. This possibility extends `awk`'s capabilities immensly. User-defined functions are called just like built-in functions, but *you* tell `awk` what to do. Thus, with user-defined functions, `awk` brings you one big step closer to becoming a smart programmer. In order to define a function you use the `function` command. It can appear anywhere in the script, since `awk` first reads the whole script and then starts to execute it. The syntax of the `function` command is:

```
function name(par1, par2,...){
  command(s)
}
```

That's it. The *name* is the name your function is to have. Here, the same rules apply as for variables and arrays. However, note that you can use any particular name only once: either as variable name, array name or function name. The *parameters* are variables you deliver to your function when you call it. Several parameters are separated by commas. The *commands* are the function's soul. They define what the function is actually supposed to do.
As usual, let us take a look at an example.

Terminal 120: Function without Parameter

```
1  $ awk 'function out() {printf "%-5s %-10s\n", $2, $1}
2  {out()}' enzyme.txt
```

```
3  Km     Enzyme
4  2.5    Protease
5  0.4    Hydrolase
6  1.2    ATPase
7  $
```

In line 1 of Terminal 120 we define a function called *out*. The function does not need any parameters, therefore the parentheses remain empty. The action of our function is to print out the positional variables *$2* and *$1* in a nicely formatted way. In line 2 we call the function, as if it were a normal command, with `out()`.
Now let us see how we can use parameters.

Terminal 121: Function with Parameters

```
1  $ awk 'function out(pre) {printf pre"%-5s %-10s\n", $2,$1}
2  {out("-> ")
3  out("# ")}' enzyme.txt
4  -> Km    Enzyme
5  # Km    Enzyme
6  -> 2.5   Protease
7  # 2.5   Protease
8  -> 0.4   Hydrolase
9  # 0.4   Hydrolase
10 -> 1.2   ATPase
11 # 1.2   ATPase
12 $
```

The function in Terminal 121 is almost the same as the function in Terminal 120. However, now we use the parameter *pre*. The value of *pre* is assigned when we call the function with `out("->␣")` in line 2. Thus, the value of *pre* becomes "->␣". In line 3 we call the function with yet another parameter. In the `printf` command in line 1 the value of *pre* is used as a prefix for the output. The result is seen in lines 4 to 11.
Up to this point we simplified the world a little bit. We have said the function `out()` in Terminal 120 does not require any parameter. This is not completely true: it uses the variables *$1* and *$2* from the script. In fact, function parameters are local variables. If we define parameters they become local variables. A local variable is a variable that is local to a function and cannot be accessed or edited outside of it, no matter whether there is another variable with the same name in script or not. On the other hand, a global variable can be accessed and edited from anywhere in the program. Thus, each function is a world for itself. The following example illustrates this.

Terminal 122: Variables in Functions

```
1  $ awk 'BEGIN{a="alpha"
2  > print "a:", a
3  > fun(); print "b:", b}
4  > function fun(){print a; b="beta"}'
```

```
5  a: alpha
6  alpha
7  b: beta
8  $
```

The function fun defined in line 4 in Terminal 122 does not declare any parameters. Thus, there are no local variables. The function does have access to the variable *a* to which we assigned a value within the script in line 1. In the same way, the script has access to the variable *b* that has been assigned in the function body in line 4. Note that it makes no difference whether you define your function at the beginning, in the middle or at the end of your script. The result will always be the same.
The next two examples demonstrate the difference between local and global variables in more detail.

Program 35: Variables in Functions

```
1  # save as fun_1.awk
2  BEGIN{
3   a="alpha";  b="beta"
4   print "a:", a; print "b:", b
5   fun("new")
6   print "a:", a; print "b:", b
7  }
8  function fun(a,b) {
9   print "fun a:", a; print "fun b:", b
10  b="BETA"
11 }
```

In Program 35 we use the two variables *a* and *b* within the script and within the function fun. However, for the function fun the variables *a* and *b* are declared as parameters and thus are local variables (line 8). Note that when we call the function in line 5 with the command fun("new"), we deliver only one parameter (new) to it. Terminal 123 shows what the program does.

Terminal 123: Variables in Functions

```
1  awk -f fun_1.awk
2  a: alpha
3  b: beta
4  fun a: new
5  fun b:
6  a: alpha
7  b: beta
```

In line 1 of Terminal 123 the program is executed. Lines 2 and 3 show the value of the variables *a* and *b*, respectively, as defined by lines 3 and 4 of Program 35. Line 5 of Program 35 calls the function fun and delivers the parameter "new" to it. This parameter is assigned to the local variable *a* of function fun. The local variable *b* of function fun remains unassigned. This

means it is empty. In line 9 of Program 35 on the page before the content of the local variables is printed. The result can be seen in lines 4 and 5 of Terminal 123 on the preceding page. As expected, local *a* has the value "new" whereas local *b* is empty. Although we assign the string "BETA" to *b* within the function `fun`, this value will neither leave the environment of the function nor change the value of global *b* ("beta"). This we check with line 6 of Program 35 on the page before. The result of this line can be seen in lines 6 and 7 of Terminal 123 on the preceding page.
Now, with the knowledge from Program 35 on the page before and Terminal 123 on the preceding page in mind, let us take a look at Program 36.

Program 36: Variables in Functions

```
1  # save as fun_2.awk
2  BEGIN{
3   a="alpha"; b="beta"
4   print "a:", a; print "b:", b
5   fun("new")
6   print "a:", a; print "b:", b
7  }
8  function fun(a) {
9   print "fun a:", a; print "fun b:", b
10  b="BETA"
11 }
```

The only difference to Program 35 on the preceding page is the declaration of parameters in line 8. The output of Program 36 is shown in Terminal 124.

Terminal 124: Variables in Functions

```
1  awk -f fun_2.awk
2  a: alpha
3  b: beta
4  fun a: new
5  fun b: beta
6  a: alpha
7  b: BETA
```

The main message of Program 36 is that *a* is a local and *b* a global variable. Thus, *b* is accessible from anywhere in the script and can be edited from everywhere in the script.
Finally, what happens if we call a function with more parameters than are declared? Then we will receive an error message because the undeclared parameter will end up nowhere.
The previous examples were educational exceptions. In larger programs you should always use local variables. Otherwise it might become really tricky to find out where a potential bug in your program lies.
With the command `return` we can specify a return value for our function. This is illustrated in the next example.

Terminal 125: return

```
$ awk '
function square(x,a){y=x**2+23x-a; return y}
{print $1" --- "$2" --- "square($2,5)}
' enzyme.txt
Enzyme --- Km --- 18
Protease --- 2.5 --- 24.25
Hydrolase --- 0.4 --- 18.16
ATPase --- 1.2 --- 19.44
$
```

The `return` command in line 1 of Terminal 125 invokes that the function call (`square($2,5)`) in line 2 returns the value of y. Alternatively, without the `return` command, you need to call the function and then state that you want to print the value of y. Of course, this works only for a global variable y. This is shown in Terminal 126.

Terminal 126: No return

```
$ awk '
function square(x,a){y=x**2+23x-a}
{print $1" --- "$2" --- "square($2,5)y}
' enzyme.txt
Enzyme --- Km --- 18
Protease --- 2.5 --- 24.25
Hydrolase --- 0.4 --- 18.16
ATPase --- 1.2 --- 19.44
$
```

What does our function `square` do? Well, it simply does some math and calculates the result of the quadratic equation from the Km values in column two of the file *enzyme.txt*. Why is the result for "Km" 18? Because `awk` does strange things when calculating with characters. Therefore, you should check whether your variable really contains numbers and no characters.

Terminal 127: Number Check

```
$ awk '
> function square(x,a){
> if (x~/[A-Z]/){
>    return ("\""x"\" no number")}
> else{
>    y=x**2+23x-a; return y}
> }
> {print $1" --- "$2" --- "square($2,5)}
> ' enzyme.txt
Enzyme --- Km --- "Km" no number
Protease --- 2.5 --- 24.25
Hydrolase --- 0.4 --- 18.16
```

```
13 ATPase --- 1.2 --- 19.44
14 $
```

Terminal 127 shows you one possible solution to this problem.

11.9 Input with `getline`

In all previous examples `awk` got its input from one or several input files. These were specified in the command line. With this technique we are restricted to access one input file at a time. With the command `getline` you can open a second input file. This feature expands `awk`'s capabilities enormously; but `getline` is even more powerful, since you can use it in order to communicate with the shell. With this section we will round up our knowledge on `awk`.
In general, `getline` reads a complete line of input (let it be from a file or the output from a shell command). Depending on the way you apply `getline`, the record is saved in one single variable or split up into fields. Furthermore, the value of different built-in variables may be influenced, in particular the positional variables (*$0* and *$1...$n*) and the "number of fields" variable (*NF*). The return value of `getline` is 1 if it read a record, 0 if it reached the end of a file or -1 if an error occurred.

11.9.1 Reading from a File

Let us assume the following scenario. You have one file containing the names of enzymes and the value of their catalytic activity. In a second file you have a list of enzyme names and the class to which they belong. Now you wish to fuse this information into one single file. How can you accomplish this task? One way is to use `getline`! In order to go through this example step by step, let us first create the desired files. The first file is actually *enzyme.txt*. You should still have it. The second file, we name it *class.txt*, has to be created.

Terminal 128: enzyme.txt and class.txt

```
1  $ cat enzyme.txt
2  Enzyme      Km
3  Protease    2.5
4  Hydrolase   0.4
5  ATPase      1.2
6  $ cat > class.txt
7  Protease = Regulation
8  ATPase = Energy
9  Hydrolase = Macro Molecules
10 Hydrogenase = Energy
11 Phosphatase = Regulation
12 $
```

Terminal 128 shows the content of the example files we are going to use. Note that both files use different field separators: in *enzyme.txt* the fields are space-delimited, whereas in *class.txt* the fields are separated by "␣=␣". Now, let us take a look at the program we are applying to solve the task.

Program 37: File Fusion

```
1  # save as enzyme-class.awk
2  # assigns a class to an enzyme
3  # requires enzyme.txt and class.txt
4  BEGIN{FS=" = "
5    while ((getline < "class.txt") > 0){
6      class[$1]=$2}
7  FS=" "
8  }
9  {if (class[$1]==""){
10   print $0 "\t\tClass"}
11 else{
12   print $0 "\t\t" class[$1]}
13 }
```

Program 37 is executed with

```
awk -f enzyme-class.awk enzyme.txt
```

The program is separated into two main parts: a `BEGIN` block (lines 4 to 8) and the main body (lines 9 to 13). In line 4 we assign the string "␣=␣" to the field separator variable *FS*. Then we use a `while` loop, spanning from line 5 to 8. This loops cycles as long as the following condition is true:

```
(getline < "class.txt") > 0
```

What does this mean? With

```
getline < "class.txt"
```

we read one line of the file *class.txt*. Note, that the "less" character (<) does not mean "less than", but is used like redirections. The complete line is saved in *$0*. Then, the lines are split according to the value of the field separator *FS* and saved in *$1...$n*, where *n* is the number of fields (which is saved in *NF*). As said before, `getline` returns 1 if it could read a line, 0 if it reached the end of the file, and -1 if it encountered an error. Thus, what we do in line 5 is to check whether `getline` read a record. In line 6 we use an array called *class* (see Sect. 11.5.4 on page 181). The enzyme name, saved in *$1*, is used as the array index, whereas the corresponding class, saved in *$2*, is used as corresponding array value. For example, the element *class["ATPase"]* has the value "Energy". Now we know what the `while` loop is doing: it reads records from the file *class.txt* and saves its fields in the array called *class*. Finally, in line 7, we set back the field separator to the space character. Now, the main body of Program 37 on the preceding page starts. We use no pattern; this means each line of the input file *enzyme.txt* is read. In line 9 we check if the value of the array element *class[$1]* is empty. For the first record of *enzyme.txt* the value of *$1* is "Enzyme". However, the array element *class["Enzyme"]* is empty because there is no corresponding line in the file *class.txt*. Thus line 8 would be executed. It prints the original record of *enzyme.txt* (saved in *$0*), plus two tabulators (`\t`), plus the word "Class". Otherwise, if the array element is not empty, *$0*, plus two tabulators, plus the value of the array element *class[$1]* is printed. The resulting output is shown in Terminal 129.

Terminal 129: File Fusion Output

```
1  $ awk -f enzyme-class.awk enzyme.txt
2  Enzyme      Km     Class
3  Protease    2.5    Regulation
4  Hydrolase   0.4    Macro Molecules
5  ATPase      1.2    Energy
6  $
```

In the previous example, the record caught by `getline` was saved in *$0* and split into the positional field variables *$1...$n*. Additionally, the variable *NF* (number of fields) was set accordingly. However, this might not be appropriate because under certain circumstances the record read from the file by `getline` would overwrite the record from the input file stated in the command line. In such cases you can save the line read by `getline` in a variable. The syntax is:

```
getline Variable < "FileName"
```

The record is then saved in *Variable* and neither *$0* nor *NF* are changed. Of course, the record is not split into fields either. However, that you could do with the `split` function (see Sect. 11.8.3 on page 196).
The following example shows you how to save a record in a variable.

Terminal 130: getline and variables

```
1  $ awk '
2  > {getline class < "class.txt"
3  > print class" <> " $0}
4  > ' enzyme.txt
5  Protease = Regulation <> Enzyme       Km
6  ATPase = Energy <> Protease     2.5
7  Hydrolase = Macro Molecules <> Hydrolase   0.4
8  Hydrogenase = Energy <> ATPase       1.2
9  $
```

The script shown in Terminal 130 reads for each line of *enzyme.txt* one line of *class.txt* and prints out both. The line from *class.txt* is saved in the variable *class*.

11.9.2 Reading from the Shell

In the previous section we used the redirection character "<" in order to read a file with `getline`. Guess what, with the pipe character "|" we can redirect the output of a command to `getline`. The syntax is

`Command | getline Variable`

The use of the variable is optional. If you do not use the variable, the record caught by `getline` is saved in *$0* and *NF* is set to the number fields saved in *$1...$n*. Splitting is executed according to the value of the field separator variable *FS*. All these variables remain unchanged, if you use a variable.
Now, let us take a look at a real-world example: we want to read the output of the shell's `date` and `pwd` commands.

Terminal 131: getline From The Shell

```
1  $ awk 'BEGIN{
2  > "date" | getline
3  > "pwd" | getline pwd
4  > print "Date & Time: " $0
5  > print "Current Directory: " pwd}'
6  Date & Time: Fre Jun 27 13:40:27 CEST 2003
7  Current Directory: /home/Freddy/awk
8  $
```

Terminal 131 shows the application of `getline` with and without a variable. In line 2, the output of the shell command `date` is saved in the positional variable *$0*, whereas line 3 assigns the output of the shell command `pwd` to the variable *pwd*.
What happens if the shell command produces a multiple-line output? Then we need to capture the lines with a `while` loop.

Terminal 132: getline With Multiple Lines

```
1  $ awk 'BEGIN{
2  > while ("ls" | getline >0){
3  >   print $0
4  > }}'
5  class.txt
6  enzyme-class.awk
7  enzyme.txt
8  genomes2.txt
9  structure.pdb
10 $
```

The shell command `ls` is a good example of a command often producing an output several lines long. Terminal 132 gives you an example of how a `while` loop can be used to catch all lines. Remember: `getline` returns 1 if a record was read. Thus, the loop between lines 2 and 4 cycles as long as there is still a record to read.

11.10 Examples

This section contains some examples which are to invite you to play around with `awk` scripts.

11.10.1 Sort an Array by Indices

With the `asort` command from Section 11.5.4 on page 181 we can sort an array according to its values but not to its indices. This small script demonstrates how to circumvent this obstacle. The script is started with

```
awk -f sort-array-elements.awk enzyme.txt
```

Note that we are using the file *enzyme.txt*.

Program 38: Sort Array Indices

```
1  # save as sort-array-elements.awk
2  # sorts an array by value of
3  # array elements
4  /[0-9]/ {data[$1]=$2; j=1}
5  END{
6  {j=1}
7  {for (i in data) {ind[j]=i; j++}}
8  {n=asort(ind)}
9  {for (i=1; i<=n; i++) {print ind[i], data[ind[i]]}}
10 }
```

The trick in Program 38 is that we generate a second array *ind* that contains the indices of the array *data*. Then *ind* is sorted and used to sort *data*. Take your time to get through and understand the program.

11.10.2 Sum up Atom Positions

The following script uses the file *structure.pdb* (see Sect. 9.1 on page 128). It sums up the decimal numbers of the atom positions. Okay, I admit that this program does not really make sense. However, it is good for learning.

Program 39: Add Atom Positions

```
1  # save as atom-sum.awk
2  # uses the file structure.pdb
3  # sums up the values in the ATOM line
4  /^ATOM/ {
5  for (i=1; i<=NF; i++){
6    if ($i ~ /[0-9].[0-9]/){
7      sum+=$i
8    }
9  }
10 print "Line "NR, "  Sum= "sum
11 }
```

You execute Program 39 with

```
awk -f atom-sum.awk structure.pdb
```

The program file and the structure file are required to be in the same directory. The `for` loop in line 5 runs for every field in the current record. The following `if` statement checks whether the current field contains a point-separated number. If so, the number is added to the variable *sum* and printed in line 10, together with the line number.

11.10.3 Translate DNA to Protein

The following program translates a DNA sequence into the corresponding protein sequence. The program is executed with

```
awk -f dna2prot.awk
```

When executed without input filename, the program accepts input from the command line. Your input will be translated when you hit (Enter). Then you can enter a new DNA sequence. To abort the program press (Ctrl)+(C). Of course, you could also pipe a sequence to the program or load a file containing a DNA sequence.

Program 40: Translate DNA

```
1  # save as dna2prot.awk
2  # translates DNA to protein
3  BEGIN{c["atg"]="MET"; c["ggc"]="THY"; c["ctg"]="CYS"}
4  {i=1; p=""}
5  {do {
6    s=substr($0, i, 3)
7    printf ("%s ", s)
8    {if (c[s]==""){p=p"    "} else {p=p c[s]" "}}
9    i=i+3}
10 while (s!="")}
11 {printf("\n%s\n", p)}
```

Let us shortly formulate the problem: the input will be a DNA sequence, the output is to be a protein sequence. Thus, we need to chop up the sequence string into triplets. This we do with the function `substr`, as shown in line 6 of Program 40. The codon is saved in the variable *s*. Then we look up if the value of *s* is in our translation table (line 8). The translation table is in an associative array called *c*. If the codon in *s* does not exist in the array *c* (thus, c[s] is empty), 4 spaces are added to the variable *p*, or else, the value of c[s] is added to *p*. We repeat this process as long as there are codons (*s* must not be empty). All the rest is nice printing. Note that we use the "`do...while`" construct here. If we started with a check if *s* is empty, the loop would never start because it is initially empty!

11.10.4 Calculate Atomic Composition of Proteins

The following program is a result of my current research. As part of this we need to know the atomic composition of a set of proteins which is saved in a FASTA formatted file. The result should be in FASTA format, too. The atomic composition of the 20 proteinogenic amino acids is available from a file called *aa_atoms.txt*. The content of this file is shown below.

aa_atoms.txt
```
Name Symbol Code Mass(-H2O) SideChain Occ.(%) S N O C H
Alanine A Ala 71.079 CH3- 7.49 0 0 0 1 3
Arginine R Arg 156.188 HN=C(NH2)-NH-(CH2)3- 5.22 0 3 0 4 10
Asparagine  N Asn 114.104 H2N-CO-CH2- 4.53 0 1 1 2 4
AsparticAcid D Asp 115.089 HOOC-CH2- 5.22 0 0 2 2 3
Cysteine C Cys 103.145 HS-CH2- 1.82 1 0 0 1 3
Glutamine Q Gln 128.131 H2N-CO-(CH2)2- 4.11 0 1 1 3 6
GlutamicAcid E Glu 129.116 HOOC-(CH2)2- 6.26 0 0 2 3 5
Glycine G Gly 57.052 H- 7.10 0 0 0 0 1
Histidine H His 137.141 N=CH-NH-CH=C-CH2- 2.23 0 2 0 4 5
Isoleucine I Ile 113.160 CH3-CH2-CH(CH3)- 5.45 0 0 0 4 9
Leucine L Leu 113.160 (CH3)2-CH-CH2- 9.06 0 0 0 4 9
Lysine K Lys 128.17 H2N-(CH2)4- 5.82 0 1 0 4 10
Methionine M Met 131.199 CH3-S-(CH2)2- 2.27 1 0 0 3 7
Phenylalanine F Phe 147.177 (C6H5)-CH2- 3.91 0 0 0 7 7
Proline  P Pro 97.117 -N-(CH2)3-CH- 5.12 0 0 0 3 6
Serine S Ser 87.078 HO-CH2- 7.34 0 0 1 1 3
Threonine T Thr 101.105 CH3-CH(OH)- 5.96 0 0 1 2 5
Tryptophan W Trp 186.213 (C6H5)-NH-CH=C-CH2- 1.32 0 1 0 9 9
Tyrosine Y Tyr 163.176 (C6H4OH)-CH2- 3.25 0 0 1 7 7
Valine V Val 99.133 CH3-CH(CH2)- 6.48 0 0 0 3 6
```

This file gives more information than we currently need. The first line specifies the content of the space-separated fields in each line. The first fields contain the full amino acid name, followed by its one-letter and three-letter code, the molecular mass excluding one water molecule (which is split off during the formation of the peptide bound), the side chain composition, its natural abundance in nature and finally the number of sulfur (S), nitrogen (N), oxygen (O), carbon (C) and hydrogen (H) atoms of the side chain. Our program first reads the information of this file and then loads the protein sequences from the FASTA file. The corresponding file name is given as command line parameter. The full code is given in the following Program.

Program 41: Protein Atomic Composition
```
# save as aacomp.awk
# calculates atomic composition of proteins
# usage: awk -f aacomp.awk protein.fasta

# ASSIGN DATA ARRAY
```

```
 6  BEGIN {
 7   while (getline < "aa_atoms.txt" > 0) {
 8     # read number of S,N,O,C,H:
 9     data[$2]=$4" " $7" " $8" " $9" " $10" " $11
10     }
11  }
12  # DECLARE FUNCTION to calculate fraction
13    function f(a,b){
14      if (a != 0 && b != 0){
15        return a/b*100
16      }
17    }
18  # PROGRAM BODY
19  {
20    if ($0 ~ /^>/) {
21      print L, MW, S, N, O, C, H, "-",
22            f(S,t), f(N,t), f(O,t), f(C,t), f(H,t)
23      print $0
24      t=0; L=0; MW=0; S=0; N=0; O=0; C=0; H=0
25    }
26    else {
27      i=1; L=L+length($0); tL=tL+length($0)
28         do {
29          aa=substr($0,i,1)
30            i++
31            # begin main
32            split(data[aa],comp) # comp[1]=mass...
33            MW=MW+comp[1]; S=S+comp[2]; N=N+comp[3]
34            O=O+comp[4]; C=C+comp[5]; H=H+comp[6]
35            tMW=tMW+comp[1]; tS=tS+comp[2]; tN=tN+comp[3]
36            tO=tO+comp[4]; tC=tC+comp[5]; tH=tH+comp[6]
37            T=tS+tN+tO+tC+tH; t=S+N+O+C+H
38            # end main
39            }
40         while (aa != "")
41    }
42  }
43  END {
44   print L, MW, S, N, O, C, H, "-",
45         f(S,t), f(N,t), f(O,t), f(C,t), f(H,t)
46   print "\n>SUMMARY"
47   print tL, tMW, tS, tN, tO, tC, tH, "-",
48         f(tS,T), f(tN,T), f(tO,T), f(tC,T), f(tH,T)
49  }
```

Okay, I agree, this is a rather large program. Let us go through it step by step. As usual, all lines beginning with the hash character are comments. In Program 41 we use quite a number of them in order to keep track of its parts.

The `BEGIN` body spans from lines 6 to 11. Here we read the data file *aa_atoms.txt*. Take a look at Section 11.9.1 on page 206 for a detailed explanation of the `while` construct. The second field of the file *aa_atoms.txt*, which is available via the variable *$2*, contains the single-letter amino acid code. This we use as index for the array *data* in line 9. Each array element consists of a space-delimited string with the following information: molecular weight and the number of sulfur, nitrogen, oxygen, carbon and hydrogen atoms of the side chain. Thus, for alanine *data* would look like:

```
data["A"] = "71.079 0 0 0 1 3"
```

In line 13 we declare a function. Remember that it does not make any difference where in your program you define functions. The function uses the two local variables *a* and *b*. The function basically returns the ratio of *a* to *b* as defined in line 15 of Program 41 on the page before.
The main body of the program spans from line 19 to 42. In line 20 we check if the current line of the input file starts with the > character. Since that line contains the identifier, it is simply printed by the `print` command in line 23. Before that, the data of the preceding protein are printed (lines 21 and 22), i.e. the length (L), molecular weight (MW), count of atoms (S, N, O, C, H), a dash and the fraction of the atom counts. This is accomplished by calling the function `f` with the number of atoms (S, N, O, C, H) and the total number of side chain atoms of the protein (t) in line 22. Since there is no protein preceding the first protein (wow, what an insight) we will get the dash only when the program passes these lines for the very first time (see line 2 of Terminal 133 on the facing page). For the same reason we check in line 13 whether any parameter handed over to the function equals zero: otherwise we might end up with a division by zero. In line 24 of Program 41 on the page before we set all counters to zero. If the current line does not start with the > character, we end up in the `else` construct starting in line 26. A local counter i is set to 1 and the number of amino acids in the current line [`length($0)`] is added to the counter variables L and tL. While the value of L reflects the length of the current protein, tL is counting the total number of amino acids in the file (this is why tL is not reset in line 24). Now we reach the kernel of the program. The `do...while` construct between lines 28 and 40 of Program 41 on the preceding page is executed as long as there are amino acids in the active line. These are getting less and less because in line 29 we extracted them one by one with the `substr` function (see also Sect. 11.8.3 on page 196). The value of the variable *aa* is the current amino acid. In line 32 we use the *data* array we generated in line 9 and apply the current amino acid (sitting in *aa*) as key. The array elements are split with the `split` function (see also Sect. 11.8.3 on page 196) into the array *comp*. The default field delimiter used by the `split` function is the space character. This is why we introduced space characters in line 9. The result of this is that *comp[1]* contains the molecular weight, *comp[2]* the number of sulfur atoms and so forth. These data we add to the corresponding variables in lines 33 to 36.

Finally, we apply the END construct to print out the data of the last protein in lines 44 and 45 (which resemble lines 21 and 22) and the cumulative data in lines 47 and 48 of Program 41 on page 213.
Now let us see how the program performs. As an input file you can use any file containing protein sequences in FASTA format. The following file, named *proteins.seq*, is one possible example:

```
proteins.seq
1  >seq1
2  MRKLVFSDTERFYRELKTALAQGEEVEVITDYERYSDLPEQLKTIFELHKNKSGMWVNV
3  TGAFIPYSFAATTINYSALYIFGGAGAGAIFGVIVGGPVGAAVGGGIGAIVGTVAVATL
4  GKHHVDIEINANGKLRFKISPSIK
5  >seq2
6  MVAQFSSSTAIAGSDSFDIRNFIDQLEPTRVKNKYICPVCGGHNLSINPNNGKYSCYNE
7  LHRDIREAIKPWTQVLEERKLGSTLSPKPLPIKAKKPATVPKVLDVDPSQLRICLLSGE
8  TTPQPVTPDFVPKSVAIRLSDSGATSQELKEIKEIEYDYGNGRKAHRFSCPCAAAPKGR
9  KTFSVSRIDPITNKVAWKKEGFWPAYRQSEAIAIIKATDGIPVLLAHEGEKCVEASRLE
10 LASITWVGSSSDRDILHSLTQIQHSTGKDFLLAYCVDNDSTGWNKQQRIKEICQQAGVS
```

The program is executed as shown in Terminal 133. Please note that lines 4 to 5 and 7 to 8 comprise usually one line of output.

```
Terminal 133: Output of aacomp.awk
1  $ awk -f aacomp.awk proteins.seq
2          -
3  >seq1
4  142 15179.5 2 38 55 411 794 - 0.153846 2.92308 4.23077
5      31.6154 61.0769
6  >seq2
7  295 32608.2 10 110 141 852 1718 - 0.353232 3.88555 4.98057
8      30.0954 60.6853
9
10 >SUMMARY
11 437 47787.7 12 148 196 1263 2512 - 0.290487 3.58267 4.74461
12     30.5737 60.8085
13 $
```

As stated above, we first get one line containing only a dash. Then follow the data for all proteins. Finally, we get the summary for all proteins. This is as if we would concatenate all proteins to one and perform the calculation. The order of the output is: protein length in amino acids, molecular weight, number of sulfur, nitrogen, oxygen, carbon and hydrogen atoms in the side chain, a dash, and the fraction of sulfur, nitrogen, oxygen, carbon and hydrogen atoms.

11.10.5 Calculate Distance Between Cysteines

Now let us look at another real-world example: the search for close cysteines in protein structure files. The scientific background of this awk script is biochemistry. The activity of some enzymes is regulated by formation or cleavage

of disulfide bridges near the protein surface. This can be catalyzed by a group of proteins known as thioredoxins. Thioredoxins are small proteins with a redox-active disulfide bridge present in the characteristic active site amino acid sequence Trp-Cys-Gly-Pro-Cys. They have a molecular mass of approximately 12 kDa and are universally distributed in animal, plant and bacterial cells. As stated above, the target enzymes, which are regulated by thioredoxin, must have two cysteine residues in close proximity to each other and to the surface. Comparison of primary structures reveals that there is no cysteine-containing consensus motif present in most of the thioredoxin-regulated target enzymes [12].
A couple of years ago we found a correlation between the concentration of reduced thioredoxin and the activity of a hydrogenase [18]. In order to identify potential cysteines that could be targets for thioredoxin, one can nowadays investigate protein structure data of crystallized hydrogenases. We just need to search for cysteine sulfur atoms that are no further than 3 Åapart. In Section 9.1 on page 128 you have already got an insight into the organization of files containing structural information. For us important lines look like this:

```
$1     $2    $3    $4   $5  $6    $7     $8      $9      $10    $11    $12
ATOM  5273  SG   CYS   L  436  7.96  9.76  73.31  0.75  14.37   S
```

The first field (*$1*) says that the line contains the information about an atom. The unique atom identity is given in the second field (*$2*). Fields *$7* to *$9* contain the x-, y-, z-coordinates of the atom, respectively. Fields *$12* and *$4* tell us that the coordinates belong to a sulfur atom (S) of a cysteine (CYS).
Okay, with this information in hand, the approach to nail down our problem should be clear!? We catch all lines where *$1* matches ATOM, *$4* matches CYS and *$12* matches S, respectively. A decent script to catch these lines is

```
awk '$1=="ATOM" && $4=="CYS" && $12=="S"' 1FRV.pdb
```

where *1FRV.pdb* is the structure file. From the matching lines we extract the coordinates and save them together with the unique atom identifier *$2*. This we could do with an array:

```
cys_x[$2]=$7; cys_y[$2]=$8; cys_z[$2]=$9
```

Finally, we calculate the distance in space from all saved sulfur atoms. We are lucky that the coordinates are given in Å; but which formula should be applied to calculate the distance? Well, do you remember your math classes?

$$d = \sqrt{(x_2 - x_1) + (y_2 - y_1) + (z_2 - z_1)}$$

This is the magic formula. A graphical representation of the coordinates of two points (atoms) in the three-dimensional space in shown in Fig. 11.1 on the facing page.

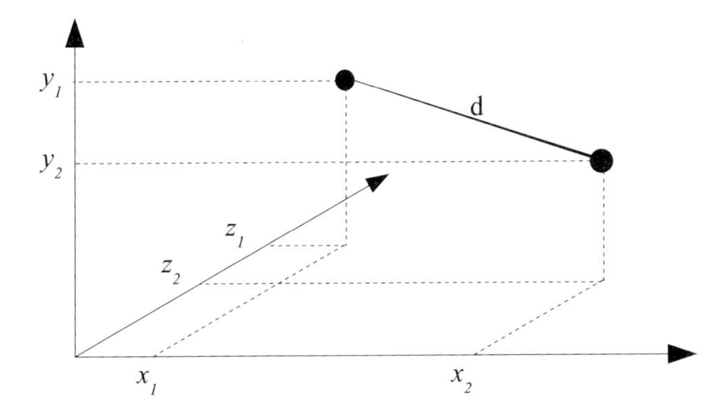

Fig. 11.1. Coordinates of two points in the three-dimensional space

Now we are missing two things: the program and the structure files. You should test two hydrogenase structures for the existence of nearby cysteines: the structure with the identifier 1FRV from the bacterium *Desulfovibrio gigas* and the structure with the identifier 1FRF from the bacterium *Desulfovibrio fructosovorans*. Both structure files can be downloaded from the Protein Data Bank at

www.rcsb.org/pdb/cgi/export.cgi/
1FRV.pdb?format=PDB"&pdbId=1FRV"&compression=None

and

www.rcsb.org/pdb/cgi/export.cgi/
1FRF.pdb?format=PDB"&pdbId=1FRV"&compression=None

respectively. You should save these files in the same directory as the `awk` script itself. Now it is time to take a look at the code in following Program.

Program 42: Cysteine Sulfur Distance

```
1  # save as distance.awk
2  # searches close cysteine sulfur atoms in a structure
3  # requires a structure file (*.pdb)
4  # usage: awk -f distance.awk structure.pdb
5
6  BEGIN{print "Cysteines in the Structure..."}
7
8  $1=="ATOM" && $4=="CYS" && $12=="S" {
9  print $4$6":", $7, $8, $9
10 cys_x[$6]=$7; cys_y[$6]=$8; cys_z[$6]=$9
11 }
12
13 END{
14 for (key1 in cys_x) {
15   for (key2 in cys_x) {
```

```
16         if (key1 != num2) {
17           dx=cys_x[key1]-cys_x[key2]
18           dy=cys_y[key1]-cys_y[key2]
19           dz=cys_z[key1]-cys_z[key2]
20           distance[key1"-"key2]=sqrt(dx^2+dy^2+dz^2)
21           if (distance[key1"-"key2]<3) {
22             i++
23             text=key1"-"key2": "distance[key1"-"key2]
24             candidate[i]=text
25           }
26         }
27       }
28     }
29     print "\nCandidates (contain doublets)..."
30     for (i in candidate) {print candidate[i]}
31     }
```

Down to line 7 nothing important is happening in Program 42. Line 8 contains the search pattern. As said above, we are looking for all lines in the protein structure file where the first field (*$1*) matches "ATOM", the fourth field (*$4*) matches "CYS" like cysteine and the last field (*$12*) matches "S" like sulfur. Note: It is very important that you place the opening brace for the action body at the end of the line containing the search pattern (line 8) and not at a new line! Otherwise the program will behave strangely. With line 9 we print out all cysteines in the file together with their unique atom ID (*$6*) and the x-, y-, z-coordinates (*$7*, *$8*, *$9*). In line 10 we use the unique atom ID as key for three arrays, *cys_x*, *cys_y*, *cys_z*, containing the coordinates. The `END` body constitutes the main part of the program performing the distance calculation. Here we apply two nested `for` loops in order to compare each atom with each atom. Remember that the `for` loop returns the array indices (or keys) in an arbitrary unordered way (see also Sect. 11.5.4 on page 183). In line 14 we pick one index from the array *cys_x* and save it in the variable *key1*. The `for` loop in line 15 picks an index from the array *cys_x*, too, and saves it in the variable *key2*. In order to avoid self-to-self comparison, we check in line 16 if the values of *key1* and *key2* are different and then perform the distance calculation in lines 17 to 20. If the distance of the current pair of sulfur atoms is less than 3 Å(line 23) apart, we add the corresponding unique atom identifiers to the array *candidate* in line 24. When you carefully analyze the current algorithm you will recognize that we compare each pair of sulfur atoms twice. Thus, we will calculate the distance of atoms A and B and later of atoms B and A. Exercise 11.5 on page 220 will deal with this problem. Finally, in lines 29 and 30 of Program 42 we print out the values of the candidate array.

Terminal 134: Cysteine Sulfur Distance

```
1  $ awk -f distance.awk 1FRF.pdb
2  Cysteines in the Structure...
3  CYS17: 22.968 5.985 85.912
```

```
4   ...
5  CYS546: 25.704 -2.660 83.522
6
7  Candidates (contain doublets)...
8  259-436: 2.55824
9  436-259: 2.55824
10 75-546: 2.41976
11 546-75: 2.41976
12 $
```

Terminal 134 shows the output of Program 42 on the facing page for the hydrogenase of the bacterium *Desulfovibrio fructosovorans* (file *1FRF.pdb*). I had to cut the output a little bit. In fact there are 20 more cysteines in the structure. Lines 8 to 11 in Terminal 134 tell us that the sulfur atoms of cysteines 259 and 436, and 75 and 546 are less than 3 Åapart. However, we do not yet know whether these sulfur atoms are close to the protein surface. To check this, we need a molecular structure viewer like Rasmol. Rasmol can be download from *openrasmol.org*. Figure 11.2 shows the structure of *Desulfovibrio fructosovorans* hydrogenase with highlighted candidate sulfur atoms.

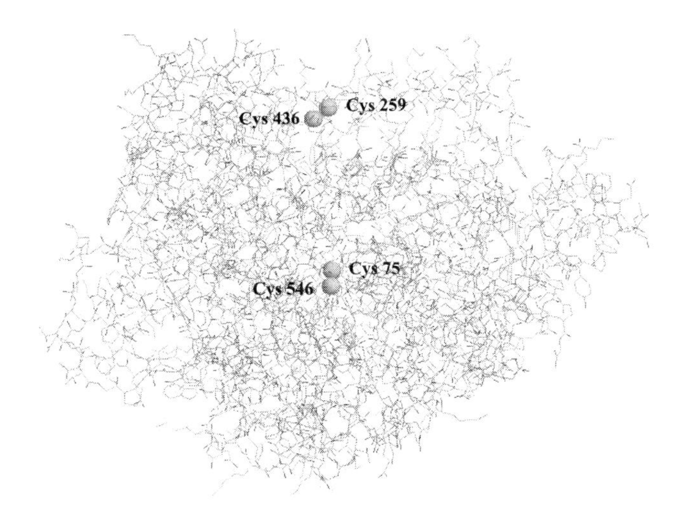

Fig. 11.2. Crystal structure of the NiFe-hydrogenase from *Desulfovibrio fructosovorans* with the candidate sulfur atoms *enlarged* and *dark*

By turning the molecule around its axis it becomes clear that only the sulfur atoms of cyteines 259 and 436 are at the surface of the protein. Thus, these are potential targets for thioredoxin. This would be the point to turn from dry science to wet science, i.e. leave the computer and do the experiment...

Exercises

Do some exercises which are hopefully not awkward. For the exercises, create two files, one containing the line "1,2,3,4,5" and another containing the line "one;two;three;four;five". Name them *numbers.txt* and *words.txt*, respectively.

11.1. Reverse the order of the numbers in *numbers.txt* and separate them with dashes (-).

11.2. Print the first line of *numbers.txt* followed by a column (:), followed by the first line of *words.txt*. Do this for all lines in both files and save the result in *number-words.txt*. The first line of the output should look like this: "`1 : one`".

11.3. Design an `awk` script that creates the following output:

```
 1  2  3  4  5  6  7  8  9 10
11 12 13 14 15 16 17 18 19 20
21 22 23 24 25 26 27 28 29 30
31 32 33 34 35 36 37 38 39 40
41 42 43 44 45 46 47 48 49 50
```

Use the `printf` command to formate the output.

11.4. In this exercise you are going to use the file *genomes2.txt* from Terminal 79 on page 167. Add one field saying how many base pairs an average gene has.

11.5. How can you avoid the display of doublets in Program 42 on page 218? Does this affect the calculation time?

12

Perl

We have to deal with perl! Why? Because it is a great and, especially among scientists, widely applied programming language – but not least also for historical reasons: perl was initially developed to integrate features of sed and awk within the framework provided by the shell. As you learned before, awk is a programming language with powerful string manipulation commands and regular expressions that facilitates file formatting and analysis. The stream editor sed nicely complements awk. In 1986, the system programmer Larry Wall worked for the US National Security Agency. He was responsible for building a control and management system with the capability to produce reports for a wide-area network of Unix computers. Unhappy with the available tools, he invented a new language: perl. Beside integrating sed and awk he was inspired by his background in linguistics to make perl a "human"-like language, which enables the expression of ideas in different ways. This feature makes perl scripts sometimes hard to read because there are too many ways to do one thing. The first version of the *Practical Extraction and Report Language*, perl, was released as open source in 1987. Since then perl has been growing and growing and growing – now even providing tools for biologists in the form of special commands via *bioperl* (see Sect. 12.13 on page 254).

12.1 Intention of this Chapter

This chapter is not intended to be a complete guide to perl. There are some really good books on this topic available, even some especially for biologists: James Tisdall [15] gives a very nice introduction to perl, whereas Rex A. Dwyer's book [4] requires some prior knowledge on programming. Both books are based on examples taken from biology. As a general introduction to perl, I liked to work with the book from Deitel et al [2]; but there are many others around! – In the course of the past chapters you have learned how to work with Unix/Linux and how to program with awk. This chapter is only going to show you the most important features of perl and its syntax. Do you remember what I said in Chapter 11 on page 163: you must learn only one programming language in order to understand all programming languages. The rest is a question of the right syntax. I can highly recommend you to buy one of these little pocket-sized *Pocket Reference* books on `perl` [17]. An old version is all you need. It provides a good compilation of all available commands and costs only around 5 Euro or US$.

12.2 Running perl Scripts

Guess what: the command in order to execute `perl` scripts is "`perl`". Like `awk`, you can either run small scripts from the command line:

```
perl -e 'print "Hello World\n"'
```

or execute perl scripts saved in a file

```
perl scriptfile.pl
```

or make a script file executable:

Terminal 135: Executable

```
1 $ cat > execute.pl
2 #!/bin/perl
3 # save as execute.pl
4 print "Hello World\n";
5 $ chmod u+x execute.pl
6 $ ./execute.pl
7 Hello World
8 $
```

In the latter case, the script file must begin with "`#!/bin/perl`". You might have noticed the differences to `awk`. There is no option necessary to execute a script file, while you need the option `-e` in order to run a command line script.
Note that each line in script files ends with the semicolon character!

12.3 Variables

In `perl`, all variables are preceded by the dollar character. The only exception is the assignment of arrays and hashes (see below). `perl` discriminates between three different variable types: *scalar variables*, *arrays* and *hashes*. Scalar variables are normal variables having either a text string or a number as values. In Section 11.5.4 on page 181 you learned what arrays are. An array variable stores a list of scalar variables. Each individual array *element* has an *index* and a *value*. The following array named x consists of 3 elements with different values:

```
x[0]="bacteria" x[1]="245.73" x[2]="\n"
```

The first element has the index "0". If the index was a text instead of a number, we are talking about associative arrays or hashes:

```
x["ATG"]="start" x["TGT"]="Cysteine" x["TGA"]="stop"
```

While `awk` did not discriminate between arrays and hashes, `perl` does, as you will see soon.
Variables are very flexible in `perl`. The value of a variable is interpreted in a certain context. Thus, the value might be interpreted either as a number or a text.

Terminal 136: Variable Context

```
1  $ perl -e '$x=7;
2  > print 1+$x,"\n";
3  > print "1+$x\n"'
4  8
5  1+7
6  $
```

In Terminal 136 we assign the value "7" to the variable called x. In the arithmetic calculation in line 2, the value of x is considered to be a number. This is also called a scalar context. In contrast, in line 3, x is considered to be a text character because it is enclosed in double quotes. It is important to keep this in mind.

12.3.1 Scalars

The names of scalar variables always begin with the dollar character ($).

Terminal 137: Scalars

```
1  $ perl -e '$x="20"; print "$x\n"'
2  20
3  $
```

Terminal 137 shows you how to assign a value to, and print the value of, a scalar variable. Of course, you can do all kinds of arithmetic calculations (+, -, *, /, **), increment (++) and decrement (--) numeric values, and use arithmetic operators for the assignment of scalar variables. The following list gives you an overview of the most common assignment operators for numeric scalar variables.

`$a += x`	Addition: `$a = $a + x`
`$a -= x`	Substraction: `$a = $a - x`
`$a *= x`	Multiplication: `$a = $a * x`
`$a /= x`	Division: `$a = $a / x`
`$a **= x`	Exponentiation: `$a = $a ** x`
`$a++`	Post-increment: increment *$a* by 1, then use the value of *$a* in the expression in which *$a* resides.
`++$a`	Pre-increment: use the current value of *$a* in the expression in which *$a* resides, then increment *$a* by 1.
`$a--`	Post-decrement: decrement *$a* by 1, then use the value of *$a* in the expression in which *$a* resides.
`--$a`	Pre-decrement: use the current value of *$a* in the expression in which *$a* resides, then decrement *$a* by 1.

In summary, apart from the dollar character preceding variable names, the handling of scalar variables is pretty much the same as in shell programming or `awk`.

12.3.2 Arrays

Arrays allow you to do almost everything with text strings and are one of the most frequently used and most powerful parts of `perl`. The name of an array variable is preceded with the vortex character (@) when assigned and when the whole array is recalled but with the dollar character ($), when a single element is recalled.

Terminal 138: Arrays

```
1  $ perl -e '@files=`ls`; print "@files"'
2  absolute.pl
3  age.pl
4  circle.pl
5  $ perl -e '@files=`ls`; print "$files[1]"'
6  age.pl
7  $
```

In the example in Terminal 138 the output of the shell command `ls` is assigned to the array named *files*. In line 1 all array elements are printed, while line 5 prints only the second array element. Remember that the first array element has the index 0.

Assigning Array Elements

There are different ways in which an array can be filled with values.

Terminal 139: Assign Array Elements
```
1  $ perl -e '$a[2]=4; print "$a[2]\n"'
2  4
3  $ perl -e '@a=(DNA, RNA, Protein); print "$a[1]\n"'
4  RNA
5  $ perl -e '@a=(DNA, RNA, Protein); print "@a[1,2]\n"'
6  RNA Protein
7  $
```

One way is to assign each element individually. This is shown in line 1 of Terminal 139. Another way is to assign a *list* to an array. A list, as shown in line 3, contains a number of comma-delimited elements enclosed in parentheses. The individual list elements are automatically assigned to array elements, starting with element 0. Individual elements can be recalled as shown in line 3, whereas line 5 demonstrates how to recall more than one element by supplying an index list (`[1,2]`).
The list to be assigned to an array must not be provided explicitly. Terminal 138 on the facing page demonstrated how the output of a shell command can be assigned to an array. Here, the output is handled as if it was a list.

Adding and Removing Elements

Once an array has been assigned, you might want to add or remove array elements. There are some very comfortable commands available for these tasks. In the following list, we assume that the array is named *array*.

`shift (@array)`
Remove First Element. – Removes and returns the first element of *array*.

`unshift (@array, list)`
Add To Beginning. – Adds the list before the first position of *array*. Returns the length of the resulting array. The list could just as well be a single string.

`pop (@array)`
Remove Last Element. – Removes and returns the last element of *array*.

`push (@array, list)`
Add To End. – Adds the list behind the last position of *array*. Returns the length of the resulting array. The list could just as well be a single string.

`@array=()`
Delete Array. – Removes all array elements by assigning an empty list to it.

`delete @array[a,b,...]`
Delete Elements. – Deletes the specified elements of *array*. Returns a list of the deleted elements.

In Terminal 140 you see some simple examples of some of the functions we have just learned.

Terminal 140: Remove and Add Elements

```
1  $ perl -e '@a=(1,2,3); print pop @a,"\t",@a,"\n"'
2  3       12
3  $ perl -e '@a=(1,2,3); print shift @a,"\t",@a,"\n"'
4  1       23
5  $ perl -e '@a=(1,2,3); unshift @a,(5,6); print "@a\n"'
6  5 6 1 2 3
7  $ perl -e '@a=(1,2,3); push @a,(5,6); print "@a\n"'
8  1 2 3 5 6
9  $ perl -e '@a=(1,2,3);
10 > print "Del: ",delete @a[1,2],"\tRest: @a\n"'
11 Del: 23 Rest: 1
12 $
```

I suppose the examples speak for themselves. In line 9 of Terminal 140 we apply a multiple-line command. Note that you need to finish each line (except the last one) with the semicolon character.

Reading Arrays

`perl` provides a very convenient command in order to read out all elements of an array.

Terminal 141: foreach

```
1  $ perl -e '@a=(one,two,three,four);
2  > foreach $e (@a){ print $e}'
3  onetwothreefour
4  $ perl -e '@a=(one,two,three,four);
5  > foreach $e (@a){ print "$e\n"}'
6  one
7  two
8  three
9  four
10 $
```

In the example shown in Terminal 141 we assign a list to the array a and use the function `foreach` to read out all array elements. `foreach` creates a loop which will be cycled until the last element of the array a is reached. The array elements are saved in the variable e.

Selecting Elements

Of course, you can use the `foreach` function in order, for example, to search for certain elements or modify individual elements that fit a search pattern; but `perl` has shortcuts for this.

`grep (condition, @array)`
The function `grep` is named after the Unix utility of the same name, which performs searches on text files (see Sect. 6.1.5 on page 68). Similarly, the `perl` function `grep` searches through an array (or a scalar) and creates a new array containing only elements that satisfy the *condition* statement.

Terminal 142: Grep Element

```
1  $ perl -e '@a=(a,1,A,aa,ab,10,5,11);
2  > @g=grep( $_ =~ /^.$/ , @a);
3  > print "@a\n@g\n"'
4  a 1 A aa ab 10 5 11
5  a 1 A 5
6  $ perl -e '@a=(a,1,A,aa,ab,10,5,11);
7  > @g=grep( $_ =~ /^[0-9]$/ , @a);
8  > print "@a\n@g\n"'
9  a 1 A aa ab 10 5 11
10 1 5
11 $
```

In Terminal 142 we create an array named *a* and use the `grep` function in line 2, to select all array elements that consist of one single character. Therefore, we apply the regular expression "`^.$`". In line 7 all array elements which have a single number as value are picked. In both examples the original array, followed by the selection, are printed. The variable *$_* is the default input, output and pattern-searching space variable (see Sect. 12.3.4 on page 233).

`map (function, @array)`
With the `map` function each element of an array can be modified.

Terminal 143: Change Elements

```
1  $ perl -e '@a=(a,1,A,aa,ab,10,5,11);
2  > @g=map( ">".$_."<" , @a);
3  > print "@a\n@g\n"'
4  a 1 A aa ab 10 5 11
5  >a< >1< >A< >aa< >ab< >10< >5< >11<
6  $
```

Terminal 143 illustrates the function of the `map` function. Again, we make use of the special variable "*$_*". The dots in line 2 are `perl`'s operators for string concatenation.

Editing Element Order

There are two important commands that affect the order of the array elements.

`reverse @array`
Reverse Element Order. – Reverses the order of the array elements. The original array remains untouched. The `reverse` function returns the new array.

Terminal 144: Reverse
```
1  $ perl -e '@a=(one,two,three,four);
2  > @r=reverse @a; print @r,"\n"'
3  fourthreetwoone
4  $ perl -e '@a=(one,two,three,four);
5  > @r=reverse @a; print "@r\n"'
6  four three two one
7  $
```

Terminal 144 demonstrates the `reverse` function. Excursion: note the effect of different quotation settings in the `print` functions in lines 2 and 5. In Section 12.6.1 on page 242, we will focus at the `print` function.

`sort {condition} @array`
Sort Elements. – The `sort` function returns a copy of the array sorted alphabetically. The original array remains unchanged. The condition, which is optional, can be set to "`$a<=>$b`" in order sort the array elements numerically.

Terminal 145: Sorting
```
1  $ perl -e '@a=(a,1,A,aa,ab,10,5,11);
2  > @alpha=sort @a;
3  > @num=sort {$a<=>$b} @a;
4  > print "@alpha\n@num\n"'
5  1 10 11 5 A a aa ab
6  a A aa ab 1 5 10 11
7  $
```

The example in Terminal 145 illustrates the application of the `sort` function. Be aware that `sort` sorts the array elements by default alphabetically.

Array Information

The following list gives you an overview of some functions in order to obtain information about an array. In the list we assume that the array is named *array*.

`scalar @array`	Returns the total number of elements of *array*.
`$#array`	Returns the index of the last array element of *array*.
`$array[-1]`	Returns the last array element's value.

Strangely enough, even when you delete an array element or assign an empty string to it, the element does still exists:

Terminal 146: scalar

```
1  $ perl -e '@a=(a,1,A,aa,ab,10,5,11);
2  > print scalar @a,"\n";
3  > delete $a[3];
4  > print scalar @a,"\n";
5  > foreach (@a){$i++};
6  > print $i,"\n";
7  > $a[4]="";
8  > print "@a\n"'
9  8
10 8
11 8
12 a 1 A   10 5 11
13 $
```

The perl script in Terminal 146 makes use of the scalar function in order to demonstrate the behaviour of an array with deleted elements: the number of array elements does not change!

12.3.3 Hashes

A hash is an unordered collection of "*index* => *value*" pairs. Hashes are also known as associative arrays, since they define associations between indices and values. While perl's arrays can only digest numeric indices, these can be either numeric or strings in hashes. The indices are sometimes referred to as *keys*. You might ask: why does perl discriminate between arrays and hashes, while awk does not (see Sect. 11.5.4 on page 181)? Well, hashes have higher memory demands. Thus, perl gives you the opportunity to be economic with your RAM (random access memory).
There is one other important thing you should know about hashes: they can be used as databases. Okay, pretty easy databases, but anyway. A hash can be exported into a file and is then available for other perl programs. We will talk in more detail about this feature later (see Sect. 12.7 on page 244).
The name of hashes is preceded with the percent character (%) when they are assigned and with the dollar character ($) when a single element is recalled.

Terminal 147: Hashes

```
1  $ perl -e '%h=(A,Adenine,C,Cytosine);print "$h{C}\n"'
2  Cytosine
3  $
```

Terminal 147 gives a short example of a hash variable. Here, the hash is created by assigning a list to it. The list is read in duplets. Thus, the first list element is used as the first index, which points to the second list element, and so forth. However, there are different methods how to create hashes and assign values to them.

Creating Hashes

In principle, hashes can be created in two ways: either by assigning a list or array to the hash (see Terminal 147 on the page before) or by assigning values to single hash elements.

Terminal 148: Create Hash

```
1 $ perl -e '$code{C}="Cys"; print "$code{C}\n"'
2 Cys
3 $ perl -e '%code=(C,Cys,A,Arg); print "$code{C}\n"'
4 Cys
5 $ perl -e '%code=(C=>Cys,A=>Arg); print "$code{C}\n"'
6 Cys
7 $ perl -e '%code=(C=>Cys,A=>Arg); print "@code{C,A}\n"'
8 Cys Arg
9 $
```

In Terminal 148 you find all three possible ways how to create a hash variable (here named *code*). In line 1 a single hash element is created by direct assignment. In lines 3 and 5 two differently formatted lists are assigned to the hash variable *code*. The last method (line 5) is probably the most readable one. Note that, in contrast to arrays, hash elements are enclosed in braces and not in brackets. Although not employed in the given examples, it is often wise to enclose text characters in single quotes, like

```
%code = ('C' => 'Cys')
```

From this example you see that `perl` is pretty much forgiving for different formats, also with respect to space characters.
Line 7 of Terminal 148 illustrates how to recall more than one hash element at once. It is important to precede the hash variable name with the vortex character (@), because you recall a list.

Hashes are not interpreted when enclosed in double quotes. This affects the action of the `print` command.

Terminal 149: Hash in Double Quotes

```
1 $ perl -e '@nuc=(a,c,g,t);
2 > %hash=(C,Cys,A,Arg);
3 > print "@nuc\n";
4 > print @nuc,"\n";
5 > print "%hash\n";
6 > print %hash,"\n"'
7 a c g t
8 acgt
```

```
9  %hash
10 AArgCCys
11 $
```

As you can see in the example in Terminal 149, arrays are interpreted both within double quotes and outside. If an array is double-quoted, the elements are separated by spaces in the output. Hashes are interpreted only outside of double quotes. Thus, the output does not look that nice. However, you can assign the space character to the output field separator variable "$,". Just add the line

```
$,=" ";
```

before the first `print` command in Terminal 149. Try it out. $, is a built-in variable. More of these are described in Section 12.3.4 on page 233.

Deleting Values

You just saw how to create hashes and add values to them. How can you delete an element?

Terminal 150: Delete Hash Element

```
1 $ perl -e '%hash=(C,Cys,A,Arg);
2 > print %hash,"\n";
3 > delete $hash{C};
4 > print %hash,"\n"'
5 AArgCCys
6 AArg
7 $
```

As with arrays, the `delete` function removes an element, here from the hash. This is illustrated in Terminal 150.

Hash Information

Of course, you can also obtain some information about hashes. The first thing you might want to know is the size of a hash variable. For this purpose you can use the `keys` function.

Terminal 151: Size of Hash

```
1 $ perl -e '%hash=`ls -l`;
2 > print eval(keys %hash),"\n"'
3 7
4 $
```

The example in Terminal 151 uses the `keys` function to evaluate the size of the hash variable called *hash*. In line 1 we use the system call `ls -l` in order to fill the hash (although the resulting hash variable does not make too much sense). We then call the keys function embedded in the eval (evaluate) function. If we omitted eval, "keys %hash" would return all indices (keys). This is shown in the next example.

Terminal 152: Keys and Values
```
1 $ perl -e '%hash=(C,Cys,A,Arg);
2 > print keys %hash,"\n";
3 > print values %hash,"\n"'
4 AC
5 ArgCys
6 $
```

In Terminal 152 we apply the functions keys and values. In a text context both return a list with all indices and all values, respectively. Remember that the indices in hashes are also called keys. Of course, you could also save the lists in an array variable instead of printing to standard output.
The keys function is very useful in order to print out all elements of a hash variable. The foreach construct in the following example resembles the one we used for arrays (see Terminal 141 on page 226).

Terminal 153: foreach
```
1 $ perl -e '%hash=(C,Cys,A,Arg); $\="-";
2 > foreach (keys %hash){print $hash{$_}}
3 > $\=""; print "\n"'
4 Arg-Cys-
5 $
```

Note that we do not specify any output variable in the foreach construct in Terminal 153. In Terminal 141 on page 226 we used the variable *$e*. If we do not specify any variable, then the default input/output variable *$_* is used. In the example in Terminal 153 we also make use of the output record separator variable *$*. By default, this is set to nothing. This is why we always use the newline escape sequence (\n) to generate a line break.

Reverse a Hash

In Section 12.3.2 on page 224 we saw how the function reverse can be applied to reverse the order of an array. If applied to hashes, reverse returns a list of pairs in which the indices and values are reversed. Thus, the indices are now values and the values are now indices.

Terminal 154: Reverse Hash
```
1 $ perl -e '%hash=(C,Cys,A,Arg);
2 > print %hash,"\n";
3 > %hash=reverse %hash;
4 > print %hash,"\n"'
5 AArgCCys
```

```
6  ArgACysC
7  $
```

The examples in Terminal 154 illustrate the effect of the `reverse` function on a hash. The `reverse` function only works correctly with hashes if all the values are unique. This is because a hash requires its keys to be unique.

12.3.4 Built-in Variables

We have already come across some built-in variables that `perl` provides for different purposes. In fact, there are over 50 built-in variables. Only some of them are listed below.

`$_`	*Default Variable.* The default input, output and pattern-searching space variable. This is the default variable for about everything.
`$/`	*Input Record Separator.* The default value is the newline character.
`$\`	*Output Record Separator.* Formats the output of the `print` command. The default value is an empty string, i.e. no record separator.
`$,`	*Output Field Separator.* Formats the output of the `print` command. The default value is an empty string, i.e. no field separator.
`$"`	*List Separator.* Formats the output of the `print` command, if an array is enclosed in double quotes. The elements are separated by the value of *$"*. The default value is the space character.
`$.`	*Input Line Number.* The current input line of the last filehandle that was read from. Reset only when the filehandle is closed explicitly.
`@ARGV`	*Command Line Parameters.* Contains a list of the command line parameters.
`%ENV`	*Environment.* Contains the shell variables. The index is the name of an environment variable, whereas the value is the current setting.

Take a look at `perl`'s manpages for more variables. Those listed above are the most commonly used ones. Most of the other variables are useful for the experienced programmer doing fancy things but not for us scientists doing basic data analysis and data formatting.

12.4 Decisions – Flow Control

I will not go through the sense of program flow controls any more. We discussed these intensively in Section 8.7 on page 110 and Section 11.7 on page 186. However, you must know the syntax of `perl`'s control structures. Furthermore, you will see that `perl` provides some additional handy flow control commands.

12.4.1 `if...elseif...else`

The syntax of the `if` structure is the same as in `awk` (see Sect. 11.7.1 on page 187), except the additional `elseif` statement. The `if` construct either executes a command if a condition is true or skips the command(s). With the help of `elseif` and `else` several cases can be distinguished.

```
if (condition-1) {
  commands-1}
elseif (condition-2) {
  command(s)-2;}
else {
  command(s)-3;}
```

There are no semicolon characters behind the braces.

12.4.2 `unless`, `die` and `warn`

The `unless` selection structure is the opposite of the `if` selection structure. `unless` either performs an action if a condition is false or skips the command(s) if the condition is true.

```
unless (condition) {
  command(s);}
```

Again, there are no semicolon characters behind the braces.
`unless` is often used in combination with the commands `die` or `warn`. `die` allows you to stop program execution and display a message, whereas `warn` displays a message without terminating program execution.

Terminal 155: die and warn

```
1  $ perl -e 'unless(1==2){die "Bye bye\n"}
2  > print "Wow: 1 is equal to 2\n";'
3  Bye bye
4  $ perl -e 'unless(1==2){warn "Bye bye\n"}
5  > print "Wow: 1 is equal to 2\n";'
6  Bye bye
7  Wow: 1 is equal to 2
8  $
```

Examples of `die` and `warn` are shown in Terminal 155. In contrast to the nonsense conditions in line 1 and 4, one would rather test for the existence of a file or something similar.

12.4.3 while...

The `while` construct performs an action while a condition is true (see Sect. 11.7.2 on page 188).

```
while (condition) {
  command(s);}
```

The `while` command first checks if the condition is true or false and then executes the command(s). There are still no semicolon characters behind the braces.

12.4.4 do...while...

Strongly related to `while`, the `do...while` construct first executes command(s) and then checks for the state of the condition (see Sect. 11.7.3 on page 189).

```
do {
  command(s);}
 while (condition)
```

There are, as usual, no semicolon characters behind the braces.

12.4.5 until...

The `until` repetition structure is simply the opposite of `while`. The body of `until` repeats while its condition is false, i.e. until its condition becomes true.

```
until (condition) {
  command(s);}
```

Note that there are no semicolon characters behind the braces.

12.4.6 do...until...

As with `do...while`, the `do...until` statement differs from the `until` statement by the time when the condition is tested with respect to command execution.

```
do {
  command(s);}
 until (condition)
```

First, the command(s) are executed, then the state of the condition is tested. Funny, but there are no semicolon characters behind the braces.

12.4.7 for...

The application of perl's for construct exactly resembles the one in awk (see Sect. 11.7.4 on page 189).

```
for (initialization; condition; counter){
  command(s);}
```

Hm, are there no semicolon characters behind the braces? No!

12.4.8 foreach...

The foreach construct allows you to iterate over a list of values, which is either provided directly or via an array.

```
foreach (list-or-array){
  command(s);}
```

There are never, ever semicolon characters behind the braces. The nice thing with explicitly given lists is that you can state ranges like "1..5" or "A..D", representing the lists "1, 2, 3, 4, 5" and "A, B, C, D", respectively.

Terminal 156: foreach

```
1  $ perl -e 'foreach $no (1..4){
2  > print $no} print "\n"'
3  1234
4  $ perl -e 'foreach (A..D){
5  > print $_} print "\n"'
6  ABCD
7  $
```

The two little scripts in Terminal 156 illustrate the use of lists. In the first example, in each cycle the values 1 to 4 are assigned to the variable *no*. In the second example the general purpose variable *$_* is used instead.
Some more examples can be found in Terminal 141 on page 226 and Terminal 142 on page 227. Especially the commands grep and map in Terminal 142 on page 227 are worth to look at.

12.4.9 Controlling Loops

Loops are nice, controlled loops are nicer. The `while`, `until`, `for` and `foreach` statements fall into the class of loop structures. Like `awk` (see Sect. 11.7.5 on page 190), `perl` allows you to control the program flow through loops.

`next` – The `next` statement skips the remaining commands in the body of a loop and initiates the next cycle (iteration) of the loop. In the following program flowchart *loop* stands for either `while`, `until`, `for` or `foreach`.

```
loop (condition){     <-
  command(s);          |
  next;               >-
  command(s);
}
```

The following example illustrates the application of `next`.

Terminal 157: next
```
1  $ perl -e '@seq=split "",$ARGV[0];
2  > foreach $nuc (@seq){
3  >   if ($nuc !~ /[acgtACGT]/){
4  >     print "-"; next;
5  >   }
6  >   print "$nuc";
7  > }
8  > print "\n"' acgTAhCsxdDVGCtacSvdg
9  acgTA-C-----GCtac---g $
```

The array element *ARGV[0]* in line 1 of Terminal 157 contains the command line parameter, here the sequence in line 8. This sequence string is split into an array (see Sect. 12.3.2 on page 224) and each element is extracted, one after the other, in the `foreach` loop. In line 3 we check if the character is a nucleotide or not. If it is not, a dash is printed in line 4 and the next cycle starts, or else, the nucleotide is printed. Note that `[acgtACGT]` is a regular expression, which must be enclosed by slashes.

`last` – The `last` command causes immediate exit from the loop. In the following program flowchart *loop* stands for either `while`, `until`, `for` or `foreach`.

```
loop (condition){
  command(s);
  last;          >-
  command(s);     |
}                 |
                 <-
```

The program execution continues with the first statement after the loop structure.

`redo` – After execution of the `redo` command the loop returns to the first command in the body of the loop without evaluating the condition. In the following program flowchart *loop* stands for either `while`, `until`, `for` or `foreach`.

```
loop (condition){
  command(s);  <-|
  redo;        >-|
  command(s);
}
```

The `redo` command is useful when it is necessary to repeat a particular cycle of a loop.

Conditions

In Section 11.4 on page 167 we learned a lot about the comparison of text strings and numbers with certain patterns. These comparisons yield either true or false and are commonly used as conditions for program flow-control structures. The list below is a compilation of the most important ones. If you can answer the question with *yes*, the condition is *true*. Usually, *$B* is not a variable but a fixed number or string.

`$A =~ /regex/`	Does *$A* match the regular expression *regex*?
`$A !~ /regex/`	Does *$A* not match the regular expression *regex*?
`$A == $B`	Is the numeric value of the *$A* equal to *$B*?
`$A eq $B`	Is the text string *$A* equal to *$B*?
`$A != $B`	Is the numeric value of the *$A* unequal to *$B*?
`$A ne $B`	Is the text string *$A* unequal to *$B*?
`$A < $B`	Is the numeric value of the *$A* smaller than *$B*? Of course, this works also for ">", "<=", and ">=".
`$A lt $B`	Is the text string *$A* less than *$B*? This works also with `gt` (greater than), `le` (less than or equal to) and `ge` (greater than or equal to). See Section 11.4.3 on page 169 for rules of string comparison.
`C1 && C2`	Are conditions *C1* and *C2* true?
`C1 || C2`	Is condition *C1* or *C2* true?
`C1 ! C2`	Is the state of condition *C1* not like the state of condition *C2*?

As I said above, there are more operators available. Those described here are the most commonly used ones.

12.5 Data Input

Well, it is nice to have a refrigerator, but it is bad if you cannot put things into it or get things out. In a sense, perl scripts and refrigerators have something in common! In this section you will see different ways which let your program read and write data.

12.5.1 Command Line

The simplest way to feed a program with data, though not the most comfortable one, is the command line. One way is to provide data in the command line when you call the program. We used this technique in Terminal 157 on page 237. The command line parameters are accessible via the array *ARGV*. The variable *ARGV[0]* is the first element in the command line and so forth. The elements are separated by space characters.
Another way is to ask the user for some input during the execution of the script.

Terminal 158: Input

```
1 $ perl -e 'print "Enter Sequence: "; $DNA=<STDIN>;
2 > print "The sequence was $DNA";
3 > print "Thank you\n"'
4 Enter Sequence: gactagtgc
5 The sequence was gactagtgc
6 Thank you
7 $
```

Terminal 158 shows you an example of how you can obtain user input. The important command is "$DNA=<STDIN>". <STDIN> is called a filehandle. You will see more of these soon. The filehandle <STDIN> initiates a request. All characters you have typed are saved in the variable *DNA* after you hit (Enter). Note that the newline character created by pressing (Enter) is saved, too. As we will see later in Section 12.9 on page 247, you can remove it with the command chomp.

12.5.2 Internal

Your program might require some default input, like a codon table. This you attach to the end of a script after the "__END__" statement.

Program 43: Internal Data

```
1 # save as data.pl
2 print <DATA>;
3 __END__
4 Here you can save your data
5 in many lines
```

In Program 43 we use the default filehandle <DATA>. With this filehandle all lines following the "__END__" statement can be recalled.

12.5.3 Files

Apart from the default filehandles <STDIN> and <DATA> you can define your own and get input from files. This is what you most probably want to do.

Program 44: Input From A File

```
1  # save as open-file.pl
2  # input a filename
3  # the content of the file is displayed
4  print "Enter Sequence File Name: ";
5  $file=<STDIN>;
6  unless (open (SEQFILE, $file)){
7  die "File does not exist\n"}
8  @content=<SEQFILE>;
9  close SEQFILE;
10 print @content;
```

Program 44 shows you how to read data from a file. This program is even a little bit interactive. Line 4 prints a text asking for a filename. Line 5 stops program execution and requires user input. After you have entered the filename, press (Enter). With the `open` command the filename saved in the variable *file* is opened. *SEQFILE* is the filehandle for this file. If you want to refer to the opened file, you have to use its filehandle. It can have any name you like. By convention, filehandles are written uppercase. We also check with the `unless...die` construct, whether the file exists. In line 8, the whole file content is assigned to the array *content*. Thus, *content[0]* will contain the first line of the file and so forth. Next, we close the file with the command `close`. Again, we use the filehandle to tell the system which file to close. This is obvious here; however, you could have many files open at once. Each file must have its own filehandle. Finally, the value of the array *content*, i.e. the file content, is displayed.
In the previous example we assigned the filehandle to an array. In such cases, the whole file is copied to the array, each line being one array element. This is a somewhat special case because usually the input file is read line by line. By default, lines (or better: records) are expected to be separated by the newline character. However, you are free to change the record separator by assigning a value to the input record separator variable *$/*. A typical file-reading procedure is shown in the next example.

Terminal 159: Read File By Lines

```
1  $ perl -e 'open (INPUT, "sequences.txt") or die;
2  > $i=1
3  > while (<INPUT>){
4  > print "Line-$i: $_";
```

```
5   > $i++}'
6   Line-1: >seq11
7   Line-2: accggttggtcc
8   Line-3: >Protein
9   Line-4: CCSTRKSBCJHBHJCBAJDHLCBH
10  $
```

In Terminal 159 the file *sequences.txt* is read line by line. The `while` loop iterates as long as there are lines to read. <INPUT> becomes false when the end of the file *sequences.txt* has been reached. The active line (record) is available via the special variable $_. Note that each line consists of all its characters plus the newline character (line break) at the end. Therefore, we do not use the newline character "\n" in the print command in line 4. At the beginning of the program we use the command `or` in conjunction with `open`. This is another way of writing `unless...`. Programmers like shortcuts!
Sometimes, you might want to read a file as blocks of characters. This can be done with the command `read`.

Terminal 160: read

```
1   $ perl -e 'open (INPUT, "sequences.txt") or die;
2   > $i=1; while (read (INPUT,$frac,10)){
3   > print "Line-$i: $frac\n";
4   > $i++}'
5   Line-1: >seq11
6   acc
7   Line-2: ggttggtcc
8   
9   Line-3: >Protein
10  C
11  Line-4: CSTRKSBCJH
12  Line-5: BHJCBAJDHL
13  Line-6: CBH
14  
15  $
```

The input file we use in Terminal 160 is the same as we used in Terminal 159. However, now we instruct `perl` only to read 10 characters from *sequences.txt* with the command

```
read (INPUT, $frac, 10)
```

The `read` command requires the filehandle, a variable name where to save the string and the number of characters to extract. Count the characters in the output by yourself and keep in mind that the invisible newline character counts as well.

12.6 Data Output

As important as data input is data output (think of the refrigerator and a cool beer). Usually, you either print your results onto the screen or save them in a file. We will deal with both cases in this section.

12.6.1 `print, printf and sprintf`

Okay, the `print` command is not really new any more. We have used it over and over in all Terminals in this chapter. Anyway, be aware that the `print` command does not automatically append a newline character. This is one important difference to `awk`. However, you can modify the default output field and record separators by changing the values of the variables `$,` and `$\`, respectively. Remember also that, in contrast to variables enclosed in single quotes, those enclosed in double quotes are expanded.
Like `awk`, `perl` offers you to print formatted text strings either onto the screen (`printf`) or into a variable (`sprintf`). Both commands work exactly as described in Section 11.8.1 on page 191 for `awk`. I am sure that you still remember how they worked, don't you?

12.6.2 Here Documents: <<

Do you recall *here documents* from shell programming? These allow you to display large blocks of text (refresh your memory in Sect. 8.4.2 on page 103). You initiate a here document with the << operator. The operator is followed by any arbitrary string (the identifier) – there must not be any space character between the operator and the identifier. All following text is regarded as coming from the standard input, until the identifier appears a second time. For `perl` to recognize the closing identifier, it must be both unquoted and at the beginning of an empty line. No other code may be placed on this line. Usually, variables within the here document are expanded. However, if the identifier is single-quoted, variables are not expanded.

Program 45: Here Document

```
# save as heredoc.pl
$text=<<TEXT;
This is line one
  this is line 2
TEXT
print <<'TEXT';
It follows the text
just saved in $text
TEXT
print "$text";
```

Program 45 demonstrates the application of here documents. The output is shown in the following Terminal.

Terminal 161: Here Documents

```
1  $ perl heredoc.pl
2  It follows the text
3  just saved in $text
4  This is line one
5    this is line 2
6  $
```

Note that the value of variable *text* in line 7 of Program 45 is not extracted because the identifier "TEXT" in line 6 is single-quoted.

12.6.3 Files

Very often one wishes to save some results directly in a file. This is basically as easy as reading a file.

Program 46: Save To File

```
1  # save as save-seq.pl
2  # appends a sequence to the file sequences.txt
3  # sequences.txt will be created if it does not exist
4  # sequences are saved in fasta format
5  print "Enter Sequence Name: ";
6  $seqname=<STDIN>;
7  print "Enter Sequence: ";
8  $seq=<STDIN>;
9  open (FILE, ">>sequences.txt");
10 print FILE ">$seqname";
11 print FILE $seq;
12 close FILE;
```

Program 46 saves a sequence together with its name in a file called *sequences.txt*. The sequence name and the sequence itself are provided via command line inputs in lines 6 and 8. In line 9 we open the file *sequences.txt* in the append mode (>>) and assign the filehandle *FILE* to it. Now, we can use the filehandle in conjunction with the `print` command in order to save data in the file *sequences.txt*. If this file does not exist, it will be created. If it exists, data will be appended to the end. Finally, in line 12, we close the file. As you could see in the previous example, files can be opened in different modes. The required mode is prefixed to the filename. The following list shows all possible modes with the corresponding prefix.

<	Open a file for reading. This is the default value.
>	Create a file for writing. If the file already exists, discard the current contents.
>>	Open or create a file for appending data at the end.
+<	Open a file for reading and writing.
+>	Create a file for reading and writing. Discard the current contents if the file already exists.
+>>	Open or create a file for reading and writing. Writing is done at the end of the file.

If you do not give any mode (as we did in Terminal 44 on page 240), the file is opened only for reading.

12.7 Hash Databases

As you have seen in Section 12.3.3 on page 229, hashes are nice tools to associate certain values. This makes you independent of zillions of `if...then` constructs. However, storage of data in variables is temporary. All such data are lost when a program terminates. `perl` provides the possibility to save hash tables as databases. These are called *DBM* files (DataBase Management). DBM files are very simple databases, though often sufficient.

Terminal 162: Write Database Files

```
1  $ perl -e '
2  unless (dbmopen(%code, "genetic-code.dbm", 0644)){
3    die "Can not open file in mode 0644\n"
4    }
5  %code=(ATG,M,TCA,S,TTC,F,TGT,C,GAT,D,TGA,"*");
6  dbmclose(%code)'
7  $
```

In Terminal 162 we create a database file with the name *genetic-code.dbm*. In line 2, this file is either opened or, if not existent, created. The command is `dbmopen`, which needs a) the hash name (*code*) preceded by a percentage character, b) the filename (*genetic-code.dbm*) enclosed in double quotes and c) the access mode (*0644*), which is the permission with which the file is to be opened (see Sect. 4.2 on page 38). In line 5 we assign a list to the hash *code* and then we close the database file. Thereby, the hash is saved in the database format. Note that the last character in the list is a special character (*), which must be enclosed in double quotes.
Now let us see how we can recall the data.

Terminal 163: Read Database File

```
1  $ perl -e '
2  unless (dbmopen(%code, "genetic-code.dbm", 0644)){
```

```
3     die "Can not open file in mode 0644\n"
4     }
5   print @code{ATG, TGT, TGT, TCA, TGA},"\n";
6   dbmclose(%code)'
7   MCCS*
8   $
```

The main structure of the program in Terminal 163 is the same as in Terminal 162 on the facing page. However, in line 5 we read out hash elements. In fact, we read several hash elements at once. Note that, in such cases, the name of the hash must be preceded with the vortex character (`@`).
That is all about `perl`'s databases. As you have seen, it is very simple and very useful to employ hash tables as databases.

12.8 Regular Expressions

Text processing is one of `perl`'s most powerful capabilities; and, as you may guess, a lot of this power comes from regular expressions. You got a thorough insight into the usage of regular expressions in Chapter 9 on page 127. Thus, I will restrict myself here to exemplify how `perl` treats regular expressions. There are some useful built-in variables that store regular expression matches. These are listed in the following table.

`$&`	The string that matched.
`` $` ``	The string preceding the matched string.
`$'`	The string following the matched string.

Of course, you can also use back referencing as described in more detail in Section 9.2.7 on page 137. Thus, *$1* would be the first parenthesized subexpression that matched, *$2* the second and so on.

12.8.1 Special Escape Sequences

In addition to the single-character meta characters we dealt with in Section 9.2.4 on page 134, `perl` interprets a number of escape sequences as query patterns. The most important ones are listed below.

`\s`	Matches space and tabulator characters.
`\S`	Inverse of `\s`
`\d`	Matches numeric characters.
`\D`	Inverse of `\d`
`\A`	Matches the beginning of a text string.
`\Z`	Matches the end of a text string.

The first four escape sequences may be used inside or outside of character classes (`[...]`).

12.8.2 Matching: m/.../

The most simple thing you might want to do is to extract a regular expression match from a string. The command would be

$var =~ m/regex/

In this case the value of the variable *var* remains unchanged and the match to *regex* would be saved in the variable *$&*. You can actually omit the command "m". Thus,

$var =~ /regex/

would do the same job. If you omit the binding operator (=~), the value of the built-in variable *$_* would be changed.
The following program gives you an example of the construct.

Program 47: Regular Expressions: match

```
1  # save as match.pl
2  while (<DATA>){
3  $text=$text.$_}
4  print "$text\n";
5  $text =~ /cc.>/s;
6  print "Prematch: $`\n";
7  print "Match: $&\n";
8  print "Postmatch: $'";
9  __END__
10 >seq11
11 accggttggtcc
12 >Protein
13 CCSTRKSBCJHBHJCBAJDHLCBH
```

In Program 47 the data is read from the end of the file. Each line is concatenated to the variable *text*. The dot in line 3 is the concatenation command. In line 4 we print out the content of *text*. Then we search for the pattern "cc.>". The modifier s tells perl to embed newline characters; thus, they can be matched with dot character (.). However, the caret (^), standing for the beginning of a line, and the dollar character ($), standing for the end of a line, no longer work in this mode. The following list gives an overview of the most important modifiers.

i *Ignore.* The query is case-insensitive.
s *Single Line.* Treat the string as a single line with embedded newline characters.
m *Multiple Line.* Treat the string as multiple lines with newline characters at the end of each line.

As with the match command m, regular expressions are used with the substitution command s and the transliteration command tr, respectively. These commands are discussed in Sections 12.9.1 and 12.9.2, respectively.

12.9 String Manipulations

As stated above, `perl` is strong in text manipulations. This is what it was developed for. Let us first see how we can substitute and transliterate certain text patterns, before we go through the most important string manipulation commands.

12.9.1 Substitute: `s/.../.../`

I guess you are pretty much accustomed to the substitution command because we have used it already a lot with `sed` and `awk`. The syntax is

```
$var =~ s/start/ATG/g
```

This would substitute all (global modifier: `g`) occurrences of "`start`" with "`ATG`" in the variable called *var*. The value of *var* is replaced by the result. If you omit the binding operator (`=~`), the value of the built-in variable *$_* would be changed.

12.9.2 Transliterate: `tr/.../.../`

The transliteration "`tr/.../.../`" resembles the "`y/.../...`" command we know from `sed` (see Sect. 10.5.2 on page 151). In fact, you can even use "`y/.../...`" – it has the same function as "`tr/.../.../`". The syntax is

```
$var =~ tr/acgt/TCGA/
```

If the value of the variable *var* was a DNA sequence, it would be complemented and converted to uppercase. Thus, "`atgcgt`" would become "`TACGCA`". If you omit the binding operator (`=~`), the value of the built-in variable *$_* would be changed.
The transliteration comes with two very useful modifiers.

`d` Deletes characters that have no corresponding replacement.
`s` Squashes duplicate consecutive matches into one replacement.

For example, the following command would replace all *a*'s with *A*'s, all *b*'s with *B*'s, and delete all *c*'s and *d*'s:

```
tr/abcd/AB/d
```

Without the modifier `d`, all *c*'s and *d*'s would be replaced by the last replacement character, here the "*B*".
An example of the application of the `s` modifier is:

```
tr/A-Z/A-Z/s
```

This would convert the word "PRRROTTEIINNN" to "PROTEIN".

12.9.3 Common Commands

In this section you will learn how to apply the most frequent text string manipulation commands. Most functions are homologous to `awk`. Thus, take a look at Section 11.8.3 on page 196 for examples.

`chomp target`
Remove Input Record Separator. – This command is usually used to remove the newline character at the end of a string. In fact, `chomp` removes the value of the input record separator variable $/ (which is by default the newline character) from the end of a string. The variable can be either a scalar or an array (then all elements are reached). The variable is modified and the number of changes is returned.

Program 48: chomp

```
1  # save as data.pl
2  while (<DATA>){
3  chomp $_;
4  print $_}
5  print "\n";
6  __END__
7  Here you can save your data
8  in many lines
```

Try out Program 48 in order to understand the function of `chomp`.

`chop target`
Remove Last Character. – This function chops off the last character of a scalar or the last element of an array. The array is deleted by this action! `chop` modifies the variable and returns the chopped character or element.

`lc target`
Lowercase. – Returns a lowercase version of *target*, which remain untouched.

`uc target`
Uppercase. – Returns an uppercase version of *target*, which remain untouched.

`length target`
Length. – Returns the length of the string *target*.

`substr(target, start, length, substitute)`
Extract Substring. – Returns a substring of *target*. The substring starts at position *start* and is of length *length*. If *length* is omitted, the rest of *target* is used. If *start* is negative, `substr` counts from the end of *target*. If *length* is negative, `substr` leaves that many characters off the end of *target*. If *substitute* is specified, the matched substring will be replaced with *substitute*.

`index(target, find, start)`
Search. – This function searches within the string *target* for the occurrence of the string *find*. The index (position from the left, given in characters) of the string *find* in the string *target* is returned (the first character in *target* is number 1). If *find* is not present, the return value is -1. If *start* is specified, `index` starts searching after the offset of "*start*" characters.

`rindex(target, find, start)`
Reverse Search. – Like `index` but returns the position of the last substring (*find*) in *target* at or before the offset *start*. Returns -1, if *find* is not found.

With the string manipulation commands presented in this section you can do basically everything. Remember that you first have to formulate clearly what your problem is. Then start to look for the right commands and play around until everything works. Very often, small errors in the use of regular expressions cause problems. You must be patient and try them out thoroughly.

12.10 Calculations

Needless to say that `perl` offers some built-in arithmetic functions to play around with. The most important ones are listed in alphabetical order below.

`abs x`	Returns the absolute value of x.
`cos x`	Returns the cosine of x in radians.
`exp x`	Returns e to the power of x.
`int x`	Returns the integer part of x.
`log x`	Returns the natural logarithm of x.
`rand x`	Returns a random fractal number between 0 (inclusive) and the value of x. The random generator can be initiated with `srand y`, where y is any number.
`sin x`	Returns the sine of x in radians.
`sqrt x`	Returns the square root of x.
`x ** y`	Raises x to the power of y.

You need the common logarithm (base 10) of x? How about your math? Some years ago? Okay, do not worry, I had to look it up, too... The logarithm of x to base b is $ln(x)/ln(b)$ where ln is the logarithm to base e. An appropriate command would be

```
$x = log($x)/log(10)
```

It would be nice to have a function for this. This we are going to learn next.

12.11 Subroutines

Imagine your programs need to calculate at many places the common logarithm of something. Instead of writing `$x = log($x)/log(10)` all the time (see Sect. 12.10 on the page before), we would be better off writing our own function. We came already across user-defined functions in `awk` (see Sect. 11.8.5 on page 201). The concept of user-defined functions is the same in `perl`; however, they are called *subroutines* and the arguments are passed differently to the subroutine. Thus, the general syntax becomes:

```
sub name{
  my($par1,$par2,...)=@_;
  command(s);
  return $whatever
}
```

You would call this subroutine with

```
name(par1,par2...)
```

The parameters, sometimes called arguments, are passed to the subroutine via the built-in array "@_". With

```
my($par1,$par2,...) = @_
```

the parameters are collected and saved in the the variables *par1*, *par2* and so on. The command `my` restricts the usage of these variables to the subroutine. You should declare all variables used in subroutines with "`my $var`" before you use them. You can also declare them while assigning them, like

```
my $var = 1.5
```

Once a variable is declared in this fashion, it exists only until the end of the subroutine. If any variable elsewhere in the program has the same name, you do not have to worry.
Okay, now back to practice. In the following example we create a subroutine that calculates the logarithm of x to the base b.

Program 49: Subroutine

```
1  # save as sublog.pl
2  # demonstrates subroutines
3  # calculates logarithm
4  print "Calculator for LOG of X to base B\n";
5  print "Enter X: "; $x=<STDIN>;
6  print "Enter B: "; $b=<STDIN>;
7  printf("%s%.4f\n", "The result is ", callog($x,$b));
8
9  sub callog{
```

```
10  my($val,$base)=@_;
11  return (log($val)/log($base));
12  }
```

The effect of the first six lines of Program 49 you should know. New is the call of the subroutine *callog* at the end of line 7. Two parameters, the variables x and b, respectively, are passed to the subroutine *callog*. The subroutine itself spans from line 9 to line 12. In line 10, the parameters are collected and saved in the variables *val* and *base*, respectively. These variables are declared with `my` and thus are valid only within the subroutine. The return value of the subroutine is the result of the calculation in line 11. Ultimately, "`callog($x,$b)`" is substituted by this return value. If necessary, you could also return a list of variables, i.e. an array. The result is formatted such that it is displayed with 4 decimal digits ("`%.4f`"; recall Sect. 11.8.1 on page 191 for more details). Terminal 164 shows the typical output of Program 49.

Terminal 164: Subroutine

```
1  $ perl sublog.pl
2  Calculator for LOG of X to base B
3  Enter X: 100
4  Enter B: 10
5  The result is 2.0000
6  $
```

If you need to, you can design your program to call subroutines from within subroutines. Sometimes, subroutines do not require any parameters; then you call them with a set of empty parentheses, like `name( )`. You can place your subroutines wherever you want in your program. However, I advise you to write them either at the beginning or at the end. This is handier, especially with large programs.

12.12 Packages and Modules

After some time of working with `perl`, you will see that there are some operations you do again and again. This could be the conversion of RNA into DNA, or the translation of RNA into a protein sequence, or calculating the common logarithm. One way to avoid writing the same program code again and again is copy-pasting, another is to include it in *packages*. A package should be saved with the file extension "*.pm*". The package itself starts with the line

```
package Name;
```

where *Name* is the package name. Package names must always begin with an uppercase character and should match the package filename! Let us stay with the logarithm problem we worked on in Section 12.11 on page 250. First, we create a package file.

Program 50: Package File

```
# save as FreddysPackage.pm
# this is a package file
package FreddysPackage;

# calculates logarithm
sub callog{
  my($val,$base)=@_;
  return (log($val)/log($base));
  }

# indicate successful import
return 1;
```

Program 50, which is a package file, contains only 2 special features. Line 3 tells `perl` that this is a package and line 12 tells the program which calls the package that everything worked fine. These two lines are necessary to define a package.
Now let us use *FreddysPackage.pm*.

Program 51: Use A Package

```
# save as require.pl
require FreddysPackage;
print "Calculator for LOG of X to base B\n";
print "Enter X: "; $x=<STDIN>;
print "Enter B: "; $b=<STDIN>;
printf("%s%.4f\n", "The result is ",
  FreddysPackage::callog($x,$b));
```

In line 2 of Program 51 we tell `perl` that we require *FreddysPackage*. Note that the file extension is omitted here. In line 7 we call the function *callog* from the package *FreddysPackage* and provide the parameters x and b. The package name and the function name are separated by the double colon operator (`::`). That's it!
After a while, it becomes boring to write the package name every time. You would rather call the subroutine *callog* from the package *FreddysPackage* as if it was a built-in function. In such cases, the package is called a module. Okay, let us go for it.

Program 52: Module

```
# save as FreddysPackage.pm
# this is a package file
package FreddysPackage;

use Exporter;
our @ISA=qw(Exporter);

```

```
 8  # export subroutine
 9  our @EXPORT=qw(&callog);
10
11  # calculates logarithm
12  sub callog{
13    my($val,$base)=@_;
14    return (log($val)/log($base));
15    }
16
17  # indicate successful import
18  return 1;
```

The package in Program 52 has more lines than Program 50 on the preceding page. Actually, I will not go into all details here. Whenever you want to make your subroutines in a package file easily accessible, add lines 5, 6 and 9. In line 9 you specify the subroutines you want to export within the parentheses, preceded with the ampersand character (&). If there are more subroutines, you have to separate them by spaces, like

```
our @EXPORT=qw(&sub1 &sub2 &sub3);
```

While the package has become more complicated, the program becomes simpler.

Program 53: Modules

```
1  # save as use.pl
2  use FreddysPackage;
3  print "Calculator for LOG of X to base B\n";
4  print "Enter X: "; $x=<STDIN>;
5  print "Enter B: "; $b=<STDIN>;
6  printf("%s%.4f\n", "The result is ",
7   callog($x,$b));
```

Instead of `require`, the command `use` is applied. In line 7 of Program 53 we call the subroutine *callog* from package *FreddysPackage* as if it was a built-in function.

With the knowledge of this section you should be able to construct your own package with the functions (modules) you frequently use. One great resource for such modules is provided by *CPAN* (Comprehensive Perl Archive Network) at *www.cpan.org*. Maybe the day will come when you place your own packages at this site.

12.13 Bioperl

Bioperl is not a new programming language but a language extension [13]. Officially organized in 1995, *The Bioperl Project* is an international association of developers of open-source `perl` tools for bioinformatics, genomics and

life-science research. Bioperl is a collection of perl modules that facilitate the development of `perl` scripts for bioinformatics applications. At the time of writing the actual bioperl version was 1.4.0 (January 2004). The newest version can be downloaded at *www.bioperl.org*. Bioperl does not include ready-to-use programs but provides reusable modules (see Sect. 12.12 on page 251) for sequence manipulation, accessing of databases using a range of data formats and execution and parsing of the results of various molecular biology programs including BLAST, Clustalw (see Chap. 5 on page 53).
Can you make use of bioperl? In principle, yes; however, in order to take advantage of bioperl, you need, apart from a basic understanding of the `perl` programming language (which you should have by now), an understanding of how to use `perl` references, modules, objects and methods. We shortly scratched at this in Section 12.12 on page 251. However, it is beyond the scope of this book to cover these topics deeply. Thus, take one of the books I recommended in the beginning, work through the chapters on references, modules, objects and methods and have fun with bioperl. A still incomplete online course can be found at *www.pasteur.fr/recherche/unites/sis/formation/bioperl*.

12.14 You Want More?

This chapter has introduced basic `perl` commands and tools. However, `perl` is much, much more powerful. It is not my intention to tell you everything - well, I do not even know everything – about programming with `perl`. With what you have learned here, you have the basic tools to solve almost all data formatting and analysis problems. If you think back to the first chapters of the book: you even had to learn how to login to and logout from the computer. Now you can even write your own programs. If you are hungry for more, I recommend you to buy a thick nice-looking book on `perl` and go ahead. There are no limits!

12.15 Examples

Let us finish this chapter with some examples.

12.15.1 Reverse Complement DNA

The following program translates a DNA sequence into an RNA sequence and computes the reverse complement of the latter.

Program 54: Transform DNA

```
1 #!/usr/bin/perl -w
2 # save as dna-rna-protein.pl
3 # playing around with sequences
```

```
4
5   print "Enter Sequence: ";
6   $DNA=<STDIN>; # ask for a seq
7   chomp $DNA; # remove end of line character
8   $DNA=uc $DNA; # make uppercase
9   print "DNA Sequence:\n$DNA\n";
10
11  $RNA = $DNA;
12  $RNA=~s/T/U/g;
13  print "RNA Sequence:\n$RNA\n";
14
15  $RevCompl=reverse $RNA;
16  $RevCompl=~tr/ACUG/UGAC/;
17  print "Reverse Complement:\n$RevCompl\n";
```

If not made executable, Program 54 is executed with

```
perl dna-rna-protein.pl
```

In the first line we see that the option `-w` is activated. This means that `perl` prints warnings about possible spelling errors and other error-prone constructs in the script. In line 5, a message is displayed that informs the user that some input is required. The DNA sequence coming from standard input (`<STDIN>`) is assigned to the variable *DNA*. In line 7, the newline character is erased with the command `chomp`. Then the sequence is made uppercase (`uc`) and displayed. In line 11, the sequence saved in *DNA* is assigned to the variable *RNA*. Then, in line 12, all *T*'s are substituted by *U*'s. The result is printed in line 13. In line 15, the value of the variable *RNA* is reversed and assigned to the variable *RevCompl*. Next, in line 16, the reversed sequence is complemented by the transliterate function "`tr/.../.../`". This is quite a nice trick to complement nucleotide sequences. Again, the result is displayed.

12.15.2 Calculate GC Content

It is a quite common problem to estimate the portion of guanine and cytosine nucleotides of a DNA sequence. The following program shows you one way in which to accomplish this task.

Program 55: GC-Content

```
1   #!/usr/bin/perl
2   # save as gccontent.pl
3   $seq = $ARGV[0];
4   print "$seq\n";
5   $gc = gc_content($seq);
6   printf("%s%.2f%s\n", "GC-Content: ",$gc,"%");
7
8   sub gc_content{
9     my $seq = shift;
```

```
10  return ($seq =~ (tr/gGcC//)/length($seq)*100);
11  }
```

Program 55 takes its input from the command line. The DNA sequence from the command line is assigned to the variable *seq* in line 3. After the sequence has been printed, the variable *seq* in line 5 is forwarded to the subroutine `gc-content`, which starts at line 8. The command `shift` in line 9 collects the first element from the array "@_" and assigns it to the local (`my`) variable *seq*. Then, in line 10, a very smart action is performed. Do you recognize what is going on? – The transliteration function is applied to erase all upper- and lowercase G's and C's. The number of replacements is divided by the length of the complete sequences and multiplied by 100, thus delivering the portion of guanines and cytosines. This line is a very good example of the versatility of `perl`. "`tr/gGcC//`" is immediately replaced by the modified string. Furthermore, since it is situated in a scalar, i.e. arithmetic, context, "`tr/gGcC//`" is replaced by its size. The result of the whole calculation is then assigned to *seq*. The result is returned and saved in *gc* in line 5. Finally, the result is printed out with two digits after the decimal point (`%.2f`).

12.15.3 Restriction Enzyme Digestion

Restriction enzymes recognize and cut DNA. The discovery of restriction enzymes laid the basis for the rise of molecular biology. Without the accurate action of restriction enzymes (or, more precisely endonucleases) gene cloning would not be possible. Those restriction enzymes that are widely used in genetic engineering recognize palindromic nucleotide sequences and cut the DNA within these sequences. Palindromes are strings that can be read from both sides, resulting in the same message. Examples are: "Madam I'm Adam" or "Sex at noon taxes".
With the following program, we wish to identify all restriction enzyme recognition sites of one or more enzymes. We do not care where the enzyme cuts (this we leave for the exercises). Furthermore, we assume that a list with restriction enzyme names and recognition sites is provided. What do we have to think of? Well, the most tricky thing will be the output. How shall we organize it, especially with respect to readability? A good solution would be to have first the original sequence, followed by the highlighted cutting sites. Try out the following program.

Program 56: Digestion

```
1  #!/bin/perl -w
2  # save as digest.pl
3  # Provide Cutter list
4  # Needs input file "cutterinput.seq" with DNA
5  %ENZYMES=(XmaIII => 'cggccg', BamHI => 'ggatcc',
6   XhoI => 'ctcgag', MstI => 'tgcgca');
7  $DNA="";
```

```
8  open (CUTTER,"cutterinput.seq") or die;
9  while (<CUTTER>){
10   chomp $_; $DNA=$DNA.$_;
11   }
12 close CUTTER;
13 $DNA = lc $DNA;
14 @OUTPUT=cutting($DNA,%ENZYMES);
15 @NAMES=keys %ENZYMES;
16 while ($DNA ne ""){
17   $i=-1;
18   printf("%-35s%s", substr($DNA,0,30,""),"\n");
19   while (++$i < scalar(@NAMES)){
20     printf("%-35s%s",
21      substr($OUTPUT[$i],0,30,""),"\t$NAMES[$i]\n");
22     }
23   }
24
25 sub cutting{
26  my ($DNA,%ENZYME)=@_;
27  foreach $ENZ (values %ENZYME){
28   @PARTS=split($ENZ,$DNA);
29   $CUT=""; $i=0;
30   while(++$i <= length($ENZ)){$CUT=$CUT.'+'}
31   foreach $NUC (@PARTS){
32    $NUC=~ s/./-/g;
33    }
34   $"=$CUT;  push(@OUT,sprintf("%-40s%s", "@PARTS","\n"));
35   }
36  chomp @OUT; return @OUT;
37 }
```

Just to give you an idea. It took me around 2 hours to write this program from scratch. The basic idea was immediately obvious: pattern matching. Formatting the output and simplifying the code took more than half of the time. Okay, let us take a look at program 56. The list of restriction enzymes and their respective cutting sites is saved in a hash variable in lines 5 and 6. The input sequences will be read from the file *cutterinput.seq*. This is done in lines 9 to 11. The input sequences are assigned to the variable *DNA*, after newline characters have been erased with `chomp`. In line 13, the sequence is converted to lowercase and then, in line 14, passed to the subroutine `cutting`. This subroutine is provided with two data sets: the DNA sequence in the variable *DNA* and the restriction enzymes plus their cutting sites in the hash variable *ENZYMES*. The subroutine starts in line 25. First, it collects the forwarded data and saves them in local variables. Then, for each (`foreach` in line 27) restriction enzymes recognition site the DNA sequences are a) cut at the recognition site (line 28), b) a string consisting of as many "+" characters as there are nucleotides in the current recognition site is created (line 30), c) all non-cut nucleotides are converted to "-" characters (line 32) and finally d) the result is appended (push) to the array *OUT* with the sprintf command. Note that the cutting site, which was converted to plus characters, is assigned to the list delimiter (*$"*). After this loop has been cycled for all restriction enzymes, the newline characters are removed from the end of the array *OUT* (line 36) and *OUT* is returned. Now we are back in line 14. The result is saved in the array *OUTPUT*. Each element of *OUTPUT* consists of the cutting pattern of *DNA* by a specific restriction enzyme. Furthermore, each element of *OUTPUT* is as long as the length of *DNA*, i.e. the DNA sequences. This is very inconvenient for the final output: with long sequences, the lines would not fit onto the screen. Thus, in line 16, we initiate a while loop that takes away portions of 30 characters of *DNA* (line 18) and *OUTPUT* (line 21). The embedded while loop from lines 19 to 22 extracts each element from *OUTPUT* and adds the name of the restriction enzyme from the array *NAMES*, which was assigned in line 15. Finally, everything is displayed with the printf command ranging from line 20 to 21.

Terminal 165: Restriction Enzymes

```
1  $ perl digest2.pl
2  tatcgatgcatcgcatgtcactagcgccgg
3  ------------------------------          XhoI
4  ------------------------------          BamHI
5  ---------------------------+++          XmaIII
6  ------------------------------          MstI
7  ccggtagtgcatcgagctagctaggatccc
8  -----------------------++++++-          XhoI
9  ------------------------------          BamHI
10 +++---------------------------          XmaIII
11 ------------------------------          MstI
12 gtcgtcgtcgtgatcgctcgagac
13 ------------------------                XhoI
14 ----------------++++++--                BamHI
15 ------------------------                XmaIII
16 ------------------------                MstI
17 $
```

Terminal 165 shows a typical output of Program 56 on the preceding page. The sequence file, containing a single sequence, must be saved in *cutterinput.seq*. All "+" characters indicate recognition sites of the corresponding restriction enzyme.

Unfortunately, Program 165 on the preceding page has a little bug: if the restriction enzyme recognition site lies at the end of the DNA sequence, the last nucleotide is not recognized. Try it out! Do you have a good idea how to cure the problem?

12.15.4 Levenshtein Distance of Sequences

This is probably the most difficult example of the whole book. Take your time and go through the example slowly and with concentration – then you will master it!
An important task in biology is to measure the difference between DNA or protein sequences. With a reasonable quantitative measure in hand we can then start to construct phylogenetic trees. Therefore, each sequence is compared with each sequence. The obtained distance measures would be filled into a so-called distance matrix, which in turn would be the basis for the construction of the tree. What is a reasonable distance measure? One way is to measure the dissimilarity between sequences as the number of editing steps to convert one sequence into the other. This resembles the action of mutations. Allowed editing operations are deletion, insertion and substitution of single characters (nucleotides or amino acids) in either sequence. Program 57 on page 261 calculates and returns the minimal number of edit operations required to change one string into another. The resulting distance measure is called Levenshtein distance. It is named after the Russian scientist Vladimir Levenshtein, who devised the algorithm in 1965 [7]. The greater the distance, the more different the strings (sequences) are. The beauty of this system is that the distance is intuitively understandable by humans and computers. Apart from biological sequence comparison there are lots of other applications of the Levenshtein distance. For example, it is used in some spell checkers to guess which word from a dictionary is meant when an unknown word is encountered.
How can we measure the Levenshtein distance? The way in which we will solve the problem to calculate the Levenshtein distance is called *dynamic programming*. Dynamic Programming refers to a very large class of algorithms. The basic idea is to break down a large problem into incremental steps so that, at any given stage, subproblems with subsolutions are obtained. Let us assume we have two sequences: ACGCTT (sequence 1) and AGCGT (sequence 2). The algorithm is based on a two-dimensional matrix (array) as illustrated in Fig. 12.1 on the next page.

The array rows are indexed by the characters of sequence 1 and the columns are indexed by the characters of sequence 2. Each cell of the array contains the number of editing steps needed to convert sequence 1 to sequence 2 at the actual cell position. In other words: each cell $[row, col]$ (row and col being the row and column index, respectively) represents the minimal distance between the first row characters of sequence 1 and the first col characters of sequence 2. Thus, cell $[2, 2]$ contains the number of editing steps needed to change AC to AG, which is 1 (convert C to G), and cell $[2, 4]$ contains the number of step to convert AC to AGCG, which is 2 (delete the Gs). How do we obtain these numbers? During an initialization step column 0 is filled with numbers ranging from 1 to the length of sequence 1 (dark shadowed in Fig. 12.1). This corresponds to the number of insertions needed if sequence 2 was zero characters long. Equally, the row 0 is filled with numbers

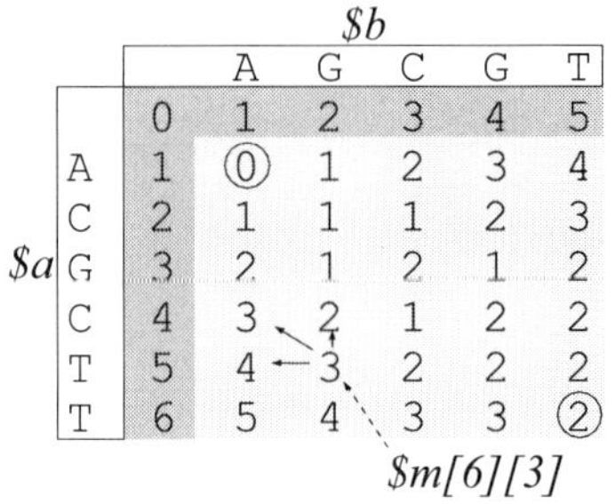

$b

		A	G	C	G	T
	0	1	2	3	4	5
A	1	(0)	1	2	3	4
C	2	1	1	1	2	3
$a G	3	2	1	2	1	2
C	4	3	2	1	2	2
T	5	4	3	2	2	2
T	6	5	4	3	3	(2)

$m[6][3]

Fig. 12.1. Construction of a matrix to calculate the Levenshtein distance between two strings. The DNA sequences, saved in variables *$a* and *$b*, respectively, are written along both sides of a matrix. The cells *underlaid in dark grey* are filled in the initialization step of the algorithm (lines 15 and 16 in Program 57 on the next page). Then, the cell values are *calculated from the top left to the bottom right* (the start and end cell are *encircled*). For example, the value of cell [6, 3], saved in the two-dimensional array *$m*, depends on the content of the upper and left three neighbours. The Levenshtein distance between the two sequences can be read directly from the bottom right cell – here the distance is 2

ranging from 1 to the length of sequence 2. This corresponds to the number of deletions if sequence 1 was zero characters long. Now, the algorithm starts in the upper left-hand corner of the array (cell [1, 1]). This cell is filled with the minimum of the following evaluation: a) the value of the cell above plus 1, b) the value of the left cell plus 1, or c) the value of the upper left diagonal cell plus 0 if the corresponding sequence characters match or plus 1 in case they do not match. Mathematically speaking, we are looking for

$$cell[row, col] = min \begin{cases} cell[row-1, col] + 1 & \\ cell[row, col-1] + 1 & \\ cell[row-1, col-1] + 1 & \text{if characters do not match} \\ cell[row-1, col-1] & \text{if characters do match} \end{cases}$$

where *row* and *col* are the current row and column, respectively. The number added to the cell value is called the *cost* for the editing step. In our example the cost for insertions, deletions and transformations is the same. For the case of sequence mutations this would correspond to the assumption that insertions, deletions and transformations are equally likely to occur, which, of course, is a simplification. Note that the cells are treated from the top left to the bottom right. A cell can only be filled when the values of all its upward and left neighbours have been filled. The bottom right cell will contain the Levenshtein distance between both sequences. Now let us take a look at the program.

Program 57: Levenshtein Distance

```
1 #!/bin/perl -w
2 # save as levenshtein.pl
```

```
3  # calculates Levenshtein distance
4  # usage: perl levenshtein.pl seq1 seq2
5
6  $one=$ARGV[0]; $two=$ARGV[1];
7  print "$one <=> $two\n";
8  print "Levenshtein Distance: ",distance($one, $two), "\n";
9
10 sub distance {
11   ($a,$b)=@_;
12   $la=length($a); $lb=length($b);
13   if(!$la) {$result=$lb;return $result}
14   if(!$lb) {$result=$la;return $result}
15   foreach $row (1 .. $la) {$m[$row][0]=$row}
16   foreach $col (1 .. $lb) {$m[0][$col]=$col}
17   foreach $row (1 .. $la) {
18     $a_i=substr($a,$row-1,1);
19     foreach $col (1 .. $lb) {
20       $b_i=substr($b,$col-1,1);
21       if ($a_i eq $b_i) {
22         $cost=0 # cost for match
23       } else {
24         $cost=1 # cost for mismatch
25       }
26       $m[$row][$col]=min($m[$row-1][$col]+1,
27                          $m[$row][$col-1]+1,
28                          $m[$row-1][$col-1]+$cost);
29     }
30   }
31   return $m[$la][$lb];
32 }
33 sub min {
34   my($a,$b,$c)=@_;
35   $result=$a;
36   if ($b < $result) {$result=$b};
37   if ($c < $result) {$result=$c};
38   return $result
39 }
```

In line 6 of Program 57 we save the sequences obtained from the command line the variables *$one* and *$two*, respectively. Therefore we employ the array *$ARGV* (see Sect. 12.3.4 on page 233). In line 8 we call the subroutine *distance* with both sequences as parameters. The subroutine itself starts in line 10. In line 11, the parameters *$one* and *$two* are saved in the variables *$a* and *@b*. Remember that the array *@_* contains the parameters the function was called with (see Sect. 12.11 on page 250). In line 12 we determine the length of *$a* and *$b* and save the values in *$la* and *$lb*, respectively. Now we check if one of the sequences is zero characters long (lines 13 and 14). In that case the Levenshtein distance equals the length of the existing sequence. In lines

15 and 16 the first row and column are initialized (see Fig. 12.1 on page 260). Take a look at Section 12.4.8 on page 236 to recall the syntax of the `foreach` construct. The `foreach` loops in lines 17 and 19 step through each row and column of the two-dimensional matrix *$m*, respectively. In lines 18 and 20 the characters corresponding to the actual cell are extracted from the sequences saved in *$a* and *$b*. If these characters match (checked in line 21), then *$cost* is set to 0, or else to 1. The construct spanning from line 26 to 28 assigns a value to the current cell $[row, col]$ saved in the array *$m*. Therefore, the subroutine *min* is applied. This subroutine, spanning from lines 33 to 39, simply returns the smallest of three numbers. What is happening in lines 26 to 28? The cost for an insertion or deletion is always 1. Thus, 1 is added to value above the current cell (*$col-1*) and left of the current cell (*$row-1*). Depending on the value of *$cost*, 1 or 0 is added to the cell diagonally above and to the left of the current cell (*row-1,col-1*). The minimum of these 3 numbers is assigned to the current cell. In this way the program surfs through the matrix. The last cell to which a value is assigned is the bottom right one (see Fig. 12.1 on page 260). Its value corresponds to the Levenshtein distance of both strings and is the return value (line 31) of the subroutine *distance*. Execution of the program is shown in Terminal 166.

Terminal 166: Levenshtein Distance

```
1 $ perl levenshtein.pl atgctatgtcgtgg tcatcgtacgtacg
2 atgctatgtcgtgg <=> tcatcgtacgtacg
3 Levenshtein Distance: 7
4 $
```

How about the complexity of our algorithm? Well, assuming that the length of each sequence is n, the running time as well as the memory demand for the matrix is n^2.
Finally, a little extension for our program that should help you to understand calculating the Levenshtein distance. Add

$$print"\backslash n";$$

to line 17 and

$$printf("\%3s", \$m[\$row][\$col]);$$

to line 28, respectively. Then the program prints out the matrix you know from Fig. 12.1 on page 260 as shown in the following Terminal.

Terminal 167: The Matrix

```
1 $ perl levenshtein2.pl at aggtgt
2 att <=> aggtgt
3
4   0  1  2  3  4  5
5   1  1  2  2  3  4
6   2  2  2  2  3  3Levenshtein Distance: 3
7 $
```

I hope you liked our excursion to dynamic programming and all the other examples! I know it is sometimes a hard job to read other people's programs, especially with `perl`; but take your time and go through the examples. This is the best way to learn `perl`. Learning is largely a process of imitating. Only when you can reproduce something can you start to modify and develop your own stuff.

Exercises

The best exercise is to apply `perl` in your daily life. However, to start with, I add some exercises referring to the examples in Section 12.15 on page 254.

12.1. Expand Program 54 on page 255 such that it translates the original DNA sequence into the corresponding protein sequence. Use the standard genetic code.

12.2. Expand Program 54 on page 255 such that it translates both the original and the complemented into the corresponding protein sequence. Use the standard genetic code.

12.3. Expand Program 54 on page 255 such that it can load several sequences in fasta format from a file and converts them.

12.4. We used Program 55 on page 256 in order to calculate the GC content of DNA. Modify it such that it accepts RNA, too.

12.5. What should be changed in Program 55 on page 256 in order to make it suitable to accept very large DNA sequences, for example from a pipe?

12.6. Modify Program 56 on page 257 such that the name of the input file can be given at the command line level.

12.7. Modify Program 56 on page 257 such that the position in terms of nucleotides is displayed at the beginning and end of each line.

12.8. Modify Program 56 on page 257 such that the list of restriction enzymes and their recognition sites is imported from a file.

12.9. Unfortunately, Program 165 on page 258 has a little bug: if the restriction enzyme recognition site lies at the end of the DNA sequence, the last nucleotide is not recognized. Try it out! Find a way to cure the problem.

A

Appendix

In this chapter you will find a number of practical hints. They are for the advanced use of Unix/Linux.

A.1 Keyboard Shortcuts

As with Windows, where you ,e.g., can terminate programs by pressing Alt+F4, there are a number of useful shortcuts in Unix/Linux, too. Here are the most common ones:

Ctrl+Alt+F1	Change to terminal 1.
Ctrl+Alt+F2	Change to terminal 2.
Ctrl+Alt+F3	Change to terminal 3.
Ctrl+Alt+F4	Change to terminal 4.
Ctrl+Alt+F5	Change to terminal 5.
Ctrl+Alt+F6	Change to terminal 6.
Ctrl+Alt+F7	Change to X-Windows.
Ctrl+ D	Logout from the active shell. If the active shell is the login shell, you logout completely from the system.
Ctrl+ C	Stop program or command execution.
Tab	Auto completion of commands and filenames in the bash shell.
↑	Recalls the last command(s) in the bash shell.
Ctrl+Alt+BkSp	Kill X-Windows.

A.2 Boot Problems

When you boot Linux and the process spends a lot of time with *sendmail* and *sm-client*, then you have a solvable problem. This phenomenon can be accompanied by the error message: "unqualified host name (localhost) unknown". Take a look at the file /etc/hosts.

Terminal 167: Boot Problem

```
1  $ cat /etc/hosts
2  # some text here
3  127.0.0.1       localhost
4  127.0.0.2       localhost.localdomain      localhost
5  $
```

If it looks like the above, or line 4 is missing, change it to:

Terminal 167: Boot Problem

```
1  $ cat /etc/hosts
2  # some text here
3  127.0.0.1       localhost
4  127.0.0.2       othername.localdomain      othername
5  $
```

A.3 Optimizing the Bash Shell

You have learned already that the bash shell is a complex tool. Let us now focus on a few advanced settings.

A.3.1 Startup Files

The bash shell uses a collection of startup files to create an environment to run in. Each file has a specific use and may affect login and interactive environments differently. These files generally sit in the */etc* directory. If an equivalent file exists in your home directory it overrides the global settings. In order to understand the different startup files, we must discriminate between different shell types. An interactive login shell is started after you successfully logged in (the leftmost shell in Fig. 7.1 on page 82). This is our command line. The interactive login shell reads the files */etc/profile* and *~/.bash_profile*. While the entries in the former affect all users, the latter file is specifically for you. Thus, this is the right place to put settings, aliases or scripts you always want to access. When you open a new shell with command `bash`, an interactive non-login shell is opened (the rightmost shells in Fig. 7.1 on page 82). It reads the files */etc/bashrc* and *~/.bashrc*. Again, the latter overrules the former. Finally, the file *~/.bash_logout* is read by the shell when you logout of the system.

A.3.2 A Helpful Shell Prompt

The following shell prompt will show your username and current directory, both in different colours:

```
PS1="\[\033]0;\w\007\033[32m\]\u@\h \[\033[33m\w\033[0m\]\n$ "
```

Do not forget the space character behind the dollar character. You see, shell prompts can be very complicated and I do not want go into details here. Write this line into *~/.bash_profile* in order to make the change permanent.

A.4 Text File Conversion (Unix ↔ DOS)

It is common that text files generated either on Unix/Linux or DOS/Windows cause problems on the other system. The reason lies in different syntax for the newline command. An easy way out is provided by the commands `unix2dos filename(s)` and `dos2unix filename(s)`, respectively.

A.5 Devices

The following list shows you the path of some more or less special devices. Unix/Linux regards all hardware as devices. They can all be found in the system folder */dev*. Remember that Unix/Linux treats everything as files, even hardware devices.

/dev/null	Efficiently disposes of unwanted output. If you save anything here, it is gone. If you read from there, the newline character is returned.
/dev/hda1	First IDE hard disk drive number (*a*), partition number (*1*). There can be several hard disk drives or partitions of a hard disk drive available. They are distinguished by their letter and number, respectively. */dev/hdc2* would be partition 2 on drive 3
/dev/sda1	First SCSI hard disk drive, partition number 1. The nomenclature is the same as for IDE hard disk drives.
/dev/fd0	The floppy disk drive.
/dev/cdrom	Guess what: this is the CD-ROM drive.

I must warn you: depending on the Unix/Linux installation you are working with, these paths might be different.

A.6 Mounting Filesystems

In Unix/Linux, filesystems like CD-ROMs are mounted. This means they look like a folder and you can place that folder whereever you want (Fig. A.1).

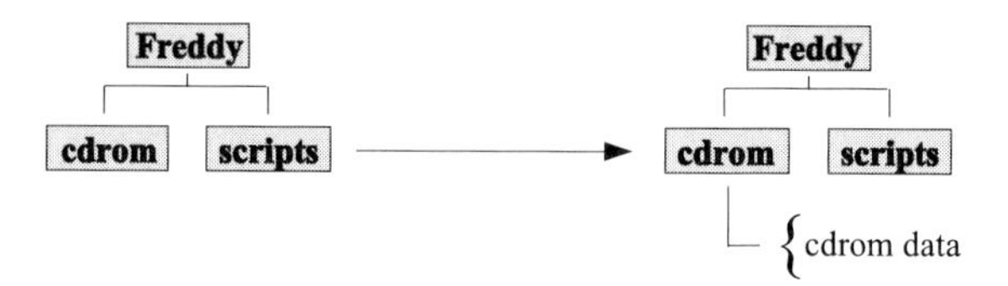

Fig. A.1. In this example the CD-ROM drive is mounted on the directory *cdrom* in Freddy's home directory. Usually, only the superuser (root) is allowed to execute the `mount` command

We have already seen in Section A.5 on the preceding page that CD-ROM drives and so on are devices. From the Windows operating system you are used to put an external memory medium like a floppy disk or CD-ROM into the drive and access it via e.g. the file manager. Depending on the settings of your Unix/Linux system you are working with, using external memory media can be more difficult. By default, Unix/Linux is protected and you can access external memory media only if the system administrator has given you the required permissions.
When you mount a filesystem, you should know what kind of filesystem it is. There several different ones around:

`ext2`	Unix/Linux filesystem (version 2).
`ext3`	Unix/Linux filesystem (version 3).
`msdos`	MS-DOS filesystem with 8+3 characters for filenames.
`vfat`	Windows 9x filesystem.
`ntfs`	Windows NT/2000/XP filesystem.
`iso9660`	CD-ROM filesystem.

In order to check which file systems and partitions are available use the `df` commando. This command reports the filesystem disk space usage.

Terminal 168: Filesystem

```
1  $ df -Th
2  Filesystem    Type    Size  Used Avail Use% Mounted on
3  /dev/hda2     ext3     18G  3.1G   13G  18% /
4  /dev/hda1     ext3     99M   14M   79M  15% /boot
5  none         tmpfs     93M  4.0K   93M   1% /dev/shm
6  $
```

In Terminal 168 we use the `df` command with the options `-T` and `-h`, which leads to the addition of the filesystem-type information and makes the output human-readable, respectively.

Now, let us see how you can add a filesystem like a floppy disk or CD-ROM drive.

A.6.1 Floppy Disk

The floppy disk drive is added to the existing filesystem with the command `mount`. Usually, you must be superuser (root) in order to be able to use `mount`. If you are working on your own system, you should have the root password. You can always login as root with the command `su`.

Terminal 169: Mount Floppy Disk

```
1  $ cat /etc/passwd | grep Freddy
2  Freddy:x:502:502::/home/Freddy:/bin/bash
3  $ echo $UID
4  502
5  $ cd
6  $ mkdir floppy
7  $ su
8  Password:
9  # mount -t auto -o uid=502 /dev/fd0 /home/Freddy/floppy/
10 # exit
11 exit
12 $ ...
13 $ cd
14 $ su
15 Password:
16 # umount /home/Freddy/floppy
17 # exit
18 exit
19 $
```

When root mounts a filesystem, it belongs to root. In order to allow you to edit files and directories, you need to know your user ID (do not confuse it with the username). Terminal 169 shows you a list of events you have to perform in order to mount a floppy disk. The first 4 lines demonstrate two possibilities to find out your user ID. Then you change to your home directory (`cd`) and create the directory *floppy*. Now, login as root (`su`) and execute the crucial command in line 9. Then, logout as root (`exit`). Now you can access and edit the floppy disk. Before you remove the floppy, unmount it! Go back into your home directory (the drive cannot be unmounted if you have a folder or file of it opened) and login as root. The correct `umount` command is shown in line 16. Now you can take out the floppy disk.

A.6.2 CD-ROM

Mounting the CD-ROM drive is pretty much like mounting a floppy-disk drive (see Sect. A.6.1 on the page before).

Terminal 170: Mount CD-ROM

```
1  $ cd
2  $ mkdir cdrom
```

```
3  $ su
4  Password:
5  #  mount -t iso9660 /dev/cdrom /home/Freddy/cdrom
6  # exit
7  exit
8  $ ...
9  $ cd
10 $ su
11 Password:
12 # umount /home/Freddy/cdrom
13 # exit
14 exit
15 $
```

Since we do not have write access to a CD-ROM anyway, we do not care about our user ID in Terminal 170. Again, note that you have to unmount the CD-ROM with `umount` before you remove it!

A.6.3 PCMCIA and Compact Flash

It is very common nowadays to have compact flash or other small memory media. I have had bad experience using USB adapters. However, with a PCMCIA adapter I have no trouble to mount my compact flash card with

```
mount -t vfat /dev/hde1 /home/Freddy/cf/
```

Tell me if you have found a nice solution for USB adapters.

A.7 Nucleotide and Protein Codes

Of course, you are welcome to learn the codes; however, you might prefer to look them up here...

Table A.1. Standard Genetic Code

1st Position	2nd Position				3rd Position
	U	C	A	G	
U	UUU Phe	UCU Ser	UAU Tyr	UGU Cys	U
	UUC Phe	UCC Ser	UAC Tyr	UGC Cys	C
	UUA Leu	UCA Ser	UAA Stop	UGA Stop	A
	UUG Leu	UCG Ser	UAG Stop	UGG Trp	G
C	CUU Leu	CCU Pro	CAU His	CGU Arg	U
	CUC Leu	CCC Pro	CAC His	CGC Arg	C
	CUA Leu	CCA Pro	CAA Gln	CGA Arg	A
	CUG Leu	CCG Pro	CAG Gln	CGG Arg	G
A	AUU Ile	ACU Thr	AAU Asn	AGU Ser	U
	AUC Ile	ACC Thr	AAC Asn	AGC Ser	C
	AUA Ile	ACA Thr	AAA Lys	AGA Arg	A
	AUG Met/Start	ACG Thr	AAG Lys	AGG Arg	G
G	GUU Val	GCU Ala	GAU Asp	GGU Gly	U
	GUC Val	GCC Ala	GAC Asp	GGC Gly	C
	GUA Val	GCA Ala	GAA Glu	GGA Gly	A
	GUG Val	GCG Ala	GAG Glu	GGG Gly	G

Next comes the single-letter and triple-letter protein code.

Table A.2. The Protein Code

Alanine	Ala	A	Cysteine	Cys	C
Aspartic Acid	Asp	D	Glutamic Acid	Glu	E
Phenylalanine	Phe	F	Glycine	Gly	G
Histidine	His	H	Isoleucine	Ile	I
Lysine	Lys	K	Leucine	Leu	L
Methionine	Met	M	Asparagine	Asn	N
Proline	Pro	P	Glutamine	Gln	Q
Arginine	Arg	R	Serine	Ser	S
Threonine	Thr	T	Valine	Val	V
Tryptophan	Trp	W	Tyrosine	Tyr	Y

A.8 Special Characters

It is quite important that you and I speak the same language when it comes to special characters like parentheses, brackets and braces. Here is a list for your orientation:

<table>
<tr><td>␣</td><td>Space</td><td>blank</td></tr>
<tr><td>#</td><td>Crosshatch</td><td>number sign, sharp, hash</td></tr>
<tr><td>$</td><td>Dollar Sign</td><td>dollar, cash, currency symbol, string</td></tr>
<tr><td>%</td><td>Percent Sign</td><td>percent, grapes</td></tr>
<tr><td>&</td><td>Ampersand</td><td>and, amper, snowman, daemon</td></tr>
<tr><td>*</td><td>Asterisk</td><td>star, spider, times, wildcard, pine cone</td></tr>
<tr><td>,</td><td>Comma</td><td>tail</td></tr>
<tr><td>.</td><td>Period</td><td>dot, decimal (point), full stop</td></tr>
<tr><td>:</td><td>Colon</td><td>two-spot, double dot, dots</td></tr>
<tr><td>;</td><td>Semicolon</td><td>semi, hybrid</td></tr>
<tr><td><></td><td>Angle Brackets</td><td>angles, funnels</td></tr>
<tr><td><</td><td>Less Than</td><td>less, read from</td></tr>
<tr><td>></td><td>Greater Than</td><td>more, write to</td></tr>
<tr><td>=</td><td>Equal Sign</td><td>equal(s),</td></tr>
<tr><td>+</td><td>Plus Sign</td><td>plus, add, cross, and, intersection</td></tr>
<tr><td>-</td><td>Dash</td><td>minus (sign), hyphen, negative (sign)</td></tr>
<tr><td>!</td><td>Exclamation Point</td><td>exclamation (mark), (ex)clam</td></tr>
<tr><td>?</td><td>Question Mark</td><td>question, query, wildchar</td></tr>
<tr><td>@</td><td>Vortex</td><td>at, each, monkey (tail)</td></tr>
<tr><td>()</td><td>Parentheses</td><td>parens, round brackets, bananas</td></tr>
<tr><td>(</td><td>Left Parentheses</td><td>open paren, wane, parenthesee</td></tr>
<tr><td>)</td><td>Right Parentheses</td><td>close paren, wax, unparenthesee</td></tr>
<tr><td>[]</td><td>Brackets</td><td>square brackets, edged parentheses</td></tr>
<tr><td>[</td><td>Left Bracket</td><td>bracket, left square bracket, opensquare</td></tr>
<tr><td>]</td><td>Right Bracket</td><td>unbracket, right square bracket, unsquare</td></tr>
<tr><td>{}</td><td>Braces</td><td>curly braces</td></tr>
<tr><td>{</td><td>Left Brace</td><td>brace, curly, leftit, embrace, openbrace</td></tr>
<tr><td>}</td><td>Right Brace</td><td>unbrace, uncurly, rytit, bracelet, close</td></tr>
<tr><td>/</td><td>Slash</td><td>stroke, diagonal, divided-by, forward slash</td></tr>
<tr><td>\</td><td>Backslash</td><td>bash, (back)slant, escape, blash</td></tr>
<tr><td>^</td><td>Circumflex</td><td>caret, top hat, cap, uphat, power</td></tr>
<tr><td>"</td><td>Double Quotes</td><td>quotation marks, literal mark, rabbit ears</td></tr>
<tr><td>'</td><td>Single Quotes</td><td>apostrophe, tick, prime</td></tr>
<tr><td>`</td><td>Grave</td><td>accent, back/left/open quote, backprime</td></tr>
<tr><td>~</td><td>Tilde</td><td>twiddle, wave, swung dash, approx</td></tr>
<tr><td>_</td><td>Underscore</td><td>underline, underbar, under, blank</td></tr>
<tr><td>|</td><td>Vertical Bar</td><td>pipe to, vertical line, broken line, bar</td></tr>
</table>

Solutions

Before you peek in here, try to solve the problems yourself. In some cases, especially for the programming exercises, you might need to spend an hour or so. Some scripts behave strangely if input files are in DOS format. In these cases use the `dos2unix filename` command to convert the file (see Sect. A.4 on page 267).

Solutions to Chapter 3

3.1 Take a look at Sect. 3.1 on page 27.

3.2 Take a look at Sect. 3.2.2 on page 32.

3.3 Type `date > the_date`, then `date >> the_date` and finally `cat the_date`

3.4 Use `passwd`

3.5 Use: `exit` or `logout` or Ctrl+D

Solutions to Chapter 4

4.1 Use: `cd` or `cd ~`; `pwd`; `ls -a`; `ls -al`

4.2 Use: `pwd`; `cd ..`; `cd /`; `cd` or `cd ~`

4.3 Use: `cd` or `cd ~`; `mkdir testdir`; `mkdir testdir/subdir`; `ls -a testdir/subdir`; `rm -r testdir/subdir`

4.4 Amazing how many directories there are, isn't it?

4.5 Use: `ls -l /`; `ls -l /etc/passwd`; probably you have all rights on */etc* and read-only right on */etc/passwd*; `cat /etc/passwd`

4.6 Use: `mkdir testdir2`; `date > testdir2/thedate.txt`; `cp -r testdir2 testdir`; `rm -r testdir2`

4.7 Use: `date > testdir/now`; `chmod 640 testdir/now` or `chmod u+rw,g+r,o-rwx testdir/now`

4.8 750=`rwxr-x---`; 600=`rw-------`; 640=`rw-r-----`; use: `chmod -R 640 testdir/*`

4.9 Use: `mkdir dirname; cd dirname`

4.10 Use: `man sort > thepage.txt`; `ls -l`; `compress thepage.txt`; `ls -l`; calculate the size ratio and multiply by 100

4.11 Use: `mkdir testpress`; `man sort > testpress/file1`; only the files, not the directory, are compressed

4.12 Use: `date > .hiddendatefile`; `ls -a`

4.13 Take a look at Sect. 4.5 on page 44.

4.14 There is no difference between moving and renaming. Copying does and moving, renaming does not change the time stamp. Check with: `ls -l`

4.15 It is increased by 1. Check with: `ls -l`

4.16 Creation and renaming does change, editing does not change the time stamp.

4.17 Have fun...

Solutions to Chapter 5

5.1 Use the same commands described in Sect. 5.1 on page 53.

5.2 Use: `gunzip tacg-3.50-src.tar.gz`; `tar xf tacg-3.50-src.tar.gz`

5.3 Use: `.configure` and then `make -j2`

5.4 Use `cat` or `less` for file viewing (see Chap. 6 on page 63).

Solutions to Chapter 6

6.1 Use: `cat > fruits.txt` and `cat >> fruits.txt`. Stop input with Ctrl+D.

6.2 Use: `cat > vegetable`; `cat fruits.txt vegetable > dinner`

6.3 Use: `sort dinner`

6.4 Take a look at Sect. 6.3 on page 72.

Solutions to Chapter 7

7.1 Use: `chsh` and later enter */bin/bash*

7.2 Use: `ps | sort +4 -n | tee filename.txt`

Solutions to Chapter 8

8.1 Hey, do not waste your time – play with the code!

Solutions to Chapter 9

9.1 I cannot help you here...

9.2 Use: `egrep '^[^␣]+␣[^␣]+␣[^␣]+$' file.txt`

9.3 Depending on your system use: `egrep -e '-[0-9]+' file.txt` or `egrep -e '\-[0-9]+' file.txt` Note the use of the '-e' switch. Without it, egrep interprets the leading minus character in the regular expression as a switch indicator.

9.4 Use: `egrep '␣-?([0-9]+\.?[0-9]*|[0-9]*\.[0-9]+)' file.txt` Here the minus is okay because it is preceded by a space character.

9.5 Use: `egrep 'ATG[ATGC]{20,}TAA' seq.file`

9.6 Use: `egrep '␣hydrogenase' file.txt`

9.7 Use: `egrep 'GT.?TACTAAC.?AG' seq.file`

9.8 Use: `egrep 'G[RT]VQGVGFR.{13}[DW]V[CN]N{3}G' sep.file` since the N stands for any amino acid you can replace all Ns by `[GPAVLIMCFYWHKRQNEDST]`

9.9 Use: `ls -l | egrep '^.{7}r'` or `ls -l | egrep '^.......r'`

Solutions to Chapter 10

10.1 Use: `sed 's/Beisel/Weisel/' structure.pdb`

10.2 Use: `sed '1,3d' structure.pdb`

10.3 Use: `sed -n '5,10p' structure.pdb` or `sed -e '1,4d' -e '11,$d' structure.pdb`

10.4 Use: `sed '/MET/d' structure.pdb`

10.5 Use: `sed -n '/HELIX.*ILE/p' structure.pdb`

10.6 Use: `sed '/^H/s/$/***/' structure.pdb`

10.7 Use: `sed '/SEQRES/s/^.*$/SEQ/' structure.pdb`

10.8 Use: `sed '/^$/d' text.file`

Solutions to Chapter 11

11.1 Use: `awk 'BEGIN{FS=","}{for (i=NF; i>0; i--){out=out$i" - "} print out}' numbers. txt`

11.2

```
BEGIN{RS=";"; i=1
  while (getline < "words.txt" >0){word[i]=$1; i++}
  RS=","}
{print $1,":",word[NR]}
```

11.3

```
BEGIN{n=-1
  for(i=1; i<51; i++){
    n++; if(n==10){print ""; n=0}; printf("%2s ",i)}
  print ""}
```

11.4 Use: `awk '{print $0, "-", $5/$8, "bp/gene"}' genomes2.txt`

11.5 Add to line 14: `list=list" "key1`; replace in line 16 `key1 != key2` with `index(list, key2)=0`; less calculation time is needed

Solutions to Chapter 12

12.1 Add the following lines at the end of the program:

```
  print "Translated Sequence:\n";
  while (length(substr($DNA,0,3)) == 3){
    $tri=substr($DNA,0,3,"");
    print aa($tri)}
  sub aa{
   my($codon)=@_;
      if ($codon =~ /GC[ATGC]/)       {return "A"} # Ala
    elsif ($codon =~ /TG[TC]/)        {return "C"} # Cys
    elsif ($codon =~ /GA[TC]/)        {return "D"} # Asp
    elsif ($codon =~ /GA[AG]/)        {return "E"} # Glu
    elsif ($codon =~ /TT[TC]/)        {return "F"} # Phe
    elsif ($codon =~ /GG[ATGC]/)      {return "G"} # Gly
    elsif ($codon =~ /CA[TC]/)        {return "H"} # His
    elsif ($codon =~ /AT[TCA]/)       {return "I"} # Ile
    elsif ($codon =~ /AA[AG]/)        {return "K"} # Lys
```

```
  elsif ($codon =~ /TT[AG]|CT[ATGC]/) {return "L"} # Leu
  elsif ($codon =~ /ATG/)             {return "M"} # Met
  elsif ($codon =~ /AA[TC]/)          {return "N"} # Asn
  elsif ($codon =~ /CC[ATGC]/)        {return "P"} # Pro
  elsif ($codon =~ /CA[AG]/)          {return "Q"} # Gln
  elsif ($codon =~ /CG[ATGC]|AG[AG]/) {return "R"} # Arg
  elsif ($codon =~ /TC[ATGC]|AG[TC]/) {return "S"} # Ser
  elsif ($codon =~ /AC[ATGC]/)        {return "T"} # Thr
  elsif ($codon =~ /GT[ATGC]/)        {return "V"} # Val
  elsif ($codon =~ /TGG/)             {return "W"} # Trp
  elsif ($codon =~ /TA[TC]/)          {return "Y"} # Tyr
  elsif ($codon =~ /TA[AG]|TGA/)      {return "*"} # Stop
  else {return "<$codon=bad-codon>"
   }
}
```

12.2 Add the following lines before the subroutine *aa* of the previous solution:

```
print "\nTranslated Complemented Sequence:\n";
$RevCompl=~tr/U/T/;
while (length(substr($RevCompl,0,3)) == 3){
 $tri=substr($RevCompl,0,3,"");
 print aa($tri)}
```

12.3 Replace the following two lines from the previous program:

```
print "Enter Sequence: ";
$DNA=<STDIN>; # ask for a seq
```

by this block:

```
print "Enter Sequence File Name: ";
$file=<STDIN>;
open (FILE, $file) or die("Could no open file...");
@content=<FILE>; close FILE;
foreach $temp (@content) {$i++; chomp $temp;
  if ($temp =~ /^>/) {$name[$i]="$temp\n";}
  else {$i--; $seq[$i]=$seq[$i].$temp}}
for ($i=1; $i < (scalar @name); $i++){
  print "\n\n$name[$i]\n"; $DNA=$seq[$i];
```

and add a closing brace "}" before the subroutine *aa*.

12.4 Headache? This makes no difference to the code... Sorry for this joke.

12.5 Do not print out the sequence again.

12.6 Replace the filename *cutterinput.seq* by the variable *$ARGV[0]*

12.7 Change the `printf` lines to:

```
printf("%4s %-30s %4s\n", $count,
   substr($DNA,0,30,""), $count+29);$count=$count+30;
printf("%s%-35s%s","      ",
   substr($OUTPUT[$i],0,30,""),"\t$NAMES[$i]\n");
```

and add `$counter=1;` at the beginning of the script.

12.8 Use a database file in order to fill the hash table as described in Sect. 12.7 on page 244.

12.9 Add `$DNA=$DNA." ";` to line 26 and replace the dot in the regular expression in line 32 by [`ACGTacgtuU`] in Program 56 on page 257.

References

1. Altschul SF, Gish W, Miller W, Myers EW Lipman DJ (1990) J Mol Biol 215:403–410
2. Deitel HM, Deitel PJ, Nieto TR, McPhie DC (2001) Perl How to Program. Prentice Hall, Upper Saddle River, NY
3. Dougherty D, Robbins A (1997) sed & awk. O'Reilly & Associates, Sebastopol, CA
4. Dwyer RA (2003) Genomic Perl. Cambridge University Press, Cambridge, New York
5. Herold H (2003) sed & awk. Addison-Wesley, Bonn Paris
6. Lamb L, Robbins A (1998) Learning the vi Editor. O'Reilly & Associates, Sebastopol, CA
7. Levenshtein V (1966) Soviet Physics Daklady, 10:707–710
8. Mangalam HJ (2002) BMC Bioinformatics 3:8
9. Qualline S (2001) Vi iMproved – VIM. New Riders Publishing, Indianapolis, IN
10. Ritchie DM, Thompson K (1974) C ACM 17:365–337
11. Robbins A (1999) VI Editor Pocket Reference. O'Reilly & Associates, Sebastopol, CA
12. Schürmann P (2000) Annu Rev Plant Physiol Plant Mol Biol 51:371–400
13. Stajich JE et al. (2002) Genome Res 12:1611–1618
14. Thompson JD, Higgins DG, Gibson TJ (1994) Nucleic Acids Research 22:4673–4680
15. Tisdall J (2001) Beginning Perl for Bioinformatics. O'Reilly & Associates, Sebastopol, CA
16. Torvalds L, Diamond D (2001) Just for Fun: The Story of an Accidental Revolutionary. HarperBusiness, NY
17. Vromans J (2000) Perl 5. O'Reilly & Associates, Sebastopol, CA
18. Wünschiers R, Heide H, Follmann H, Senger H, Schulz R (1999) FEBS Lett 455:162–164

Index

Commands for shell programming, `awk` or `perl` will be found at these entries.